Springer Theses

Recognizing Outstanding Ph.D. Research

Aims and Scope

The series "Springer Theses" brings together a selection of the very best Ph.D. theses from around the world and across the physical sciences. Nominated and endorsed by two recognized specialists, each published volume has been selected for its scientific excellence and the high impact of its contents for the pertinent field of research. For greater accessibility to non-specialists, the published versions include an extended introduction, as well as a foreword by the student's supervisor explaining the special relevance of the work for the field. As a whole, the series will provide a valuable resource both for newcomers to the research fields described, and for other scientists seeking detailed background information on special questions. Finally, it provides an accredited documentation of the valuable contributions made by today's younger generation of scientists.

Theses are accepted into the series by invited nomination only and must fulfill all of the following criteria

- They must be written in good English.
- The topic should fall within the confines of Chemistry, Physics, Earth Sciences, Engineering and related interdisciplinary fields such as Materials, Nanoscience, Chemical Engineering, Complex Systems and Biophysics.
- The work reported in the thesis must represent a significant scientific advance.
- If the thesis includes previously published material, permission to reproduce this must be gained from the respective copyright holder.
- They must have been examined and passed during the 12 months prior to nomination.
- Each thesis should include a foreword by the supervisor outlining the significance of its content.
- The theses should have a clearly defined structure including an introduction accessible to scientists not expert in that particular field.

More information about this series at http://www.springer.com/series/8790

Thi Ha Kyaw

Towards a Scalable Quantum Computing Platform in the Ultrastrong Coupling Regime

Doctoral Thesis accepted by
the National University of Singapore,
Singapore

 Springer

Author
Dr. Thi Ha Kyaw
Bahen Centre for Information Technology
University of Toronto
Toronto, ON, Canada

Supervisor
Prof. Leong-Chuan Kwek
Centre for Quantum Technologies
National University of Singapore
Singapore

ISSN 2190-5053   ISSN 2190-5061   (electronic)
Springer Theses
ISBN 978-3-030-19660-8   ISBN 978-3-030-19658-5   (eBook)
https://doi.org/10.1007/978-3-030-19658-5

This Springer imprint is published by the registered company Springer Nature Switzerland AG
The registered company address is: Gewerbestrasse 11, 6330 Cham, Switzerland

For my parents,
Khoo Soo Khoon & Tan Shu War ...

Where there's a will, there's a way, kind of beautiful
And every night has its day, so magical
And if there's love in this life, there's no obstacle
That can't be defeated...

       —Avicii (Waiting for Love)

Supervisor's Foreword

A superconducting quantum circuit is essentially a nonlinear inductor coupled to a capacitor. The inductor is a nonlinear component called a Josephson junction. This combination effectively forms the artificial atom or the superconducting qubit. Indeed, the superconducting circuit has long been favored as one of the major platforms for the realization of a large-scale quantum processor. Advances in superconducting qubit architecture have been fueled in part by the promise of possible scalability through lithography fabrication, improvement in coherence times, and the flexibility of electrical control with standard microwave and radio-frequency techniques.

This thesis presents a neat yet brief introduction to the basics of superconducting circuits. Realizations of superconducting architecture typically lies within the weak coupling regime, where the atom–cavity coupling is small compared to the atomic or cavity frequencies. However, stronger coupling can lead to highly entangled states, a feature that is desirable for quantum processors. In this thesis, Thi Ha describes how one can describe light–matter interaction in the ultrastrong regime through the quantum Rabi model.

Despite recent great advances in the field of superconducting circuits, the field is still replete with technological challenges. In fact, two of the most pressing problems are quantum error correction and the design of a quantum memory. The author has made a good attempt at proposing solutions to these problems in the ultrastrong coupling regime in this thesis. Admittedly, these proposals are just the first steps in these directions. Finally, the author proposes a simple catalytic quantum Rabi model that can be used for quantum transport between qubits.

Finally, I must add that this thesis contains several original results and scholarly works. The thesis offers new directions that need to be explored and readers of this thesis can certainly gain significant insights into some of the problems presented in the thesis.

Singapore   Leong-Chuan Kwek
March 2019

Abstract

Superconducting circuit platform is at present a prime candidate toward realizing a practical quantum computer. Besides the good controllability and scalability features, the architecture also allows us to explore an unexplored physics territory, namely the ultrastrong light–matter coupling/interaction (USC), where the light–matter interaction strength is comparable to or even larger than the cavity and qubit frequencies. Specifically, the standard light–matter interaction in quantum optics experiment is about 10^{-6} as compared to the cavity frequency. However, due to the strong field confinement present in the superconducting chip, the light–matter interaction can also increase up to 1.34 of the cavity frequency. In this regime, the usual Jaynes–Cummings model of light–matter interaction is no longer applicable. A more general approach known as the quantum Rabi model is needed.

In this thesis, we present all our investigations focusing on the superconducting circuit architecture operating at the USC regime. The ultrastrong coupling naturally provides ultrafast two-qubit gates with which we propose to construct large quantum graph codes. We are able to create a cluster state within a fraction of a nanosecond by creating entanglement between any pair of qubits. To exemplify our proposal, creations of the five-qubit and Steane codes are numerically simulated. We also provide optimal operating conditions with which the two graph codes can be realized. Besides the ultrafast gate, the quantum Rabi model has built-in $\mathbb{Z}_2$ parity symmetry. We then move on to propose parity-protected quantum memory, whose coherence lifetime can be three orders of magnitude longer than conventional quantum memory in the strong light–matter coupling limit. We are also convinced that our proposal might pave a way to realize a scalable quantum random-access memory due to its fast storage and readout performances. Lastly, we study the catalytic quantum Rabi model where the system is composed of strong–ultrastrong coupling light–matter hybrid. By tuning the energy gap of the qubits while keeping the ultrastrong coupling system in its ground state, we demonstrate a strong two-qubit interaction as well as an enhanced excitation transfer between the two qubits. Our proposal has twofold implications: a means to attain multipurpose parity-protected quantum information tasks in superconducting circuits and a building block for ultrastrong coupled cavity-enhanced exciton transport in disordered media.

Parts of this thesis have been published in the following journal articles:

1. **Parity-preserving light-matter system mediates effective two-body interactions**.
   T. H. Kyaw, S. Allende, L.-C. Kwek and G. Romero, *Quantum Sci. Technol.* **2**, 025007 (2017).

2. **Scalable quantum memory in the ultrastrong coupling regime**.
   T. H. Kyaw, S. Felicetti, G. Romero, E. Solano and L.-C. Kwek, *Sci. Rep.* **5**, 8621 (2015).

3. **Creation of quantum error correcting codes in the ultrastrong coupling regime**.
   T. H. Kyaw, D. Herrera-Martí, E. Solano, G. Romero and L.-C. Kwek, *Phys. Rev. B* **91**, 064503 (2015).

4. **$\mathbb{Z}_2$ quantum memory implemented on circuit quantum electrodynamics**.
   T. H. Kyaw, S. Felicetti, G. Romero, E. Solano and L.-C. Kwek, *Proc. SPIE.* **9225**, 92250B (2014).

Acknowledgements

My life at Centre for Quantum Technologies (CQT) would not have been such a joyful, fulfilled, complete, satisfied, and enriched one, without the following individuals. It is my utmost blessing and good fortune to have my thesis supervisor, Prof. Leong-Chuan Kwek. Without him, my life trajectory would have been a complete difference. His encouragement, dedication, and support are unparalleled, and for that, I am very much indebted. Undoubtedly, all our group members enjoy very much of "Kwek-ing",[1] in one way or another no matter how busy or stressful we were. Thanks a lot, Kwek! I also like to thank my main thesis advisory committee member, Prof. Berthold-Georg Englert, for his guidance and support throughout my entire stay at CQT.

I am very glad to have met Guillermo Romero, who turns out to be one of my best buddies. I enjoyed numerous discussions about physics in general, superconducting circuits in particular and many other things about life with him. I appreciate our long and persistent arguments on quantum simulations, which spanned over the course of about 4 years have finally come to a mutual agreement and satisfaction. What an epic! I have learned a great deal about superconducting circuits architecture from Yiming Wang, Simone Felicetti, and Klaus Mølmer, and I am grateful to have met them. I am also very thankful to Benoît Grémaud, George Batrouni, Victor M. Bastidas, and Luigi Amico, for revealing so many cool stuff about the condensed matter physics. I learned so much about the measurement-based quantum computing, and the quantum error correction, from Ying Li, Yicong Zheng, and David A. Herrera-Martí. I always enjoyed talking with Sai Vinjanampathy, who literally knows almost everything from Indian army arsenals to gravitational wave detectors. The conversations with him were always enriching. But, I still don't understand why he dislikes so much about black holes! I greatly admire Andy Chia's meticulousness, which I will try myself very hard to level up in the near future. And, his little course on the open quantum systems during our group meeting was one of the best. I thank my friends Adolfo del Campo and Daniel

---

[1]Kwek (noun) is a Chinese surname. In Pinyin, it is written as Guō. Kwek (verb) is the action of Leong-Chuan Kwek or actions associated to Teddy bear, Ted, Paddington, Winnie-the-Pooh, etc.

Terno for inspiring me in the fields of shortcut to adiabaticity and relativistic quantum information.

I am very grateful for the hospitality I had during my QUTIS group visit in Bilbao. For that, I thank Enrique Solano, Lucas Lamata, and Jorge Casanova. My USACH, Chile visits would not be that joyful and productive, without Guillermo Romero, Sebastian Allende, Juan Carlos Retamal, Francisco Andrés Cárdenas, and Felipe Herrera. I have learned so much about experimental circuit QED from visiting NTT Basic Research Laboratories. And, I am very glad to receive the hospitality from Shiro Saito, Yuichiro Matsuzaki, and the entire crew of the USC experiment. Furthermore, I like to thank Mykola Bordyuh and Hakan Türeci, for hosting me during my visit to Princeton. Lastly, I acknowledge Yuimaru Kubo and Daniel Esteve for their kind hospitality during my visit to the Quantronics group, CEA-Saclay.

I am grateful that I shared the office/floor with my group mates: Chang Jian Kwong, Ewan Munro, Jiang Zhang, Wei Nie, Davit Aghamalyan, Andy Chia, Kishor Bharti, Tobias Haug, Hermanni Heimonen, and Shabnam Safaei. Our regular lunches and dinners were awesome. The most memorable one was when we were all at the Pangkil island beach, enjoying our seclusive annual workshop. I enjoyed discussing physics and talking everything under the sky with all of them. And, thank you all for being such an excellent company. Life was good with each and everyone, in this humble little room #04-14C at CQT.

I like to thank Leong-Chuan Kwek and Guillermo Romero for their dedication and support in correcting this thesis even though they are heavily occupied with grant proposals, research, and teaching duties. In addition, I like to thank my three thesis examiners: Prof. Wenhui Li, Prof. Maciej Lewenstein, and Prof. Siyuan Han, for providing me invaluable feedback and suggestions to improve the quality of this thesis that the reader is reading now. I also like to thank the entire CQT administrative team especially Evon Tan, Siew Hoon Lim, Chui Theng Chan, and Ai Leng Irene Tan, for helping me with all the administrative matters. I am grateful for the help received from Gowtham Chakravarthy, Shalini Selvam and Angela Lahee of Springer during the publication process.

Last but not least, it is my utmost pleasure and gratitude to have my parents' unfailing love and support through ups and downs over all these years to pursue my dream. There is no word to express how grateful I am. It is definitely not an easy task for them to raise my sister and me up. I am very glad they have done that all these whiles. Also, I am blessed to have my little sister, who always gives me headaches, but at the same time, teaches me how to be more human in many ways. It is my priceless fortune to have my best friend, soul mate, and wife, Jumiati Wu, for her boundless dedication, love, and sacrifice. Without her constant support and encouragement, this thesis would not have been completed. I dedicate this thesis to my family.

Toronto, Canada   Thi Ha Kyaw
March 2019

Contents

1   **Introduction**   1
    1.1   Structure of the Thesis   3
    References   4

2   **Basics of Superconducting Circuits Architecture**   7
    2.1   An LC Oscillator   7
    2.2   Circuit Quantization   10
       2.2.1   An Example   10
    2.3   Transmission Line   13
    2.4   Resonators   15
    2.5   Josephson Junction   17
       2.5.1   Coherent Phenomena in Superconductivity   17
       2.5.2   Josephson Effect   19
       2.5.3   Energy Storage   20
       2.5.4   Nonlinear Inductance   20
       2.5.5   Josephson Oscillations   20
       2.5.6   Other Current Components   21
    2.6   DC-SQUID   21
    2.7   Single Cooper-Pair Box and Transmon   23
    2.8   Flux Qubit   25
    2.9   Mendeleev-Like Table for Superconducting Qubits   30
    References   30

3   **Ultrastrong Light–Matter Interaction**   33
    3.1   Cavity Quantum Electrodynamics   33
       3.1.1   Strong Light–Matter Interaction   35
    3.2   Circuit Implementation of Cavity QED   38
    3.3   Quantum Rabi Model   39
    3.4   Rounding Up   43
    References   44

4   Quantum Error-Correcting Codes in the USC Regime 47
   4.1   Pairwise Cluster State Generation 47
      4.1.1   Five-Qubit Code 55
      4.1.2   Steane Code 57
   4.2   Errors and Decoherence Model 58
      4.2.1   Nonzero Transversal Light–Matter Coupling 58
      4.2.2   Decoherence Noise Modeling via Monte Carlo
            Simulation 59
   4.3   Summary and Discussions 61
   References 62

5   Quantum Memory in the USC Regime 65
   5.1   Quantum Memory Cell 66
   5.2   Generating and Catching Flying Qubits 68
   5.3   Storage and Retrieval of Flying Qubits 69
      5.3.1   The Storage Protocol 69
      5.3.2   The Retrieval Protocol 70
      5.3.3   Unavoidable Phase Imprinting During Storage
            and Retrieval 70
      5.3.4   Storage and Retrieval of Entangled Flying Qubits 72
      5.3.5   Open Quantum System Treatment 72
   5.4   Scaling up to Two Dimensions 74
      5.4.1   Cavity Network 74
   5.5   Summary and Discussions 76
   References 76

6   Catalytic Quantum Rabi Model 79
   6.1   The Simple Model 80
   6.2   Equilibrium and Nonequilibrium Properties 81
      6.2.1   Effective Two-Body Interaction in Dispersive Limit 81
      6.2.2   Entangling Two Qubits 87
      6.2.3   Excitation Transfer Between Qubits 89
   6.3   Transmon-Based Implementation 93
   6.4   Summary and Discussions 95
   References 95

7   Conclusion and Future Work 99
   7.1   Conclusion 99
   7.2   Future Work 101
   References 103

Appendix A: Derivation of the USC Evolution Operator.   105

**Appendix B: Microscopic Derivation of an Open Quantum
           System in the USC Regime**.   107

**Appendix C: Derivation of the Effective Hamiltonian
           via the Schrieffer–Wolff Transformation**.   111

Chapter 1
Introduction

The first step in solving a problem is to recognize that it does exist.

—Zig Ziglar

Since its idealistic proposal to build a computing machine using quantum mechanics by the aspiring physicist, Richard P. Feynman, in the 1980s [1], the term "quantum computer" becomes a holy grail of many physicists, computer scientists, and engineers, because it promises speedup and robust computational power over its classical counterparts. Since then, many incredible and outstanding progress were made in the field of quantum computation [2]. Through inventions of Shor factoring algorithm [3] and Grover search algorithm [4], as well as Bennett et.al.'s [5–7], Ekert's [8] quantum key distribution (QKD) protocols, quantum computing was not merely theorists' playground. It had since then attracted many outstanding scientists at various national university research laboratories, institutes, agencies, and multinational technology companies.

With the recent achievements in the first demonstration of intercontinental QKD [9] by the Chinese Academy of Sciences in Beijing and the Austrian Academy of Sciences in Vienna, as well as the fabrication of good-quality superconducting qubits by Google and IBM, we are already well into the second wave of quantum technology revolution. The benefits of achieving a practical quantum computer are far-reaching that many companies such as Intel, Microsoft, Alibaba, and Tencent are starting to enter the quantum computing race. Also, there are now many quantum computing startups focusing on quantum hardware and software. At present, we are living in the exciting era of quantum computing research frontier [10].

However, anyone who is familiar with quantum computing research would clearly know that a practical quantum computer is still far from reality despite the numerous great achievements. One major challenge simply arises from the fact that the qubits realized in the laboratories are prone to external noises. Scaling up to many good-

© Springer Nature Switzerland AG 2019
T. H. Kyaw, *Towards a Scalable Quantum Computing Platform in the Ultrastrong Coupling Regime*, Springer Theses,
https://doi.org/10.1007/978-3-030-19658-5_1

quality qubits without generating unwanted noise is a very daunting task. When we use the word "many", we desire hundreds, or probably thousands in number. As this thesis is being written, the current record for the largest number of qubits manufactured on a superconducting chip is 72 demonstrated by Google (Bristlecone design), and was announced during the APS March Meeting 2018. Google team is also a leader in the manufacture of good-quality qubits as good as 99.92% fidelity for single-qubit gate and 99.4% fidelity for two-qubit gate [11]. Even with the high level of controllability [11–13] and scalability features [14, 15] provided by the superconducting circuit platform, we are still far away from an ultimate practical use since the numbers (qubits, gates fidelity) achieved so far are still not good enough. In fact, as recently developed by IBM Q team, it is not the number of qubits present in a quantum processor a key factor rather its "Quantum Volume" [16]. It is a measurement developed by IBM that determines how powerful a quantum computer is, after taking into both gate and measurement errors, device cross talk, as well as device connectivity and circuit compiler efficiency. The higher the Quantum Volume, the more real-world, complex problems quantum computers can potentially tackle. In this aspect, IBM achieves highest Quantum Volume, which was announced during the APS March meeting 2019, and it also establishes a roadmap for reaching quantum advantage by doubling the Quantum Volume every year.

Besides the superconducting circuits, there are also some other promising quantum computing platforms in the market—namely trapped ions, nitrogen-vacancy centers in diamond, linear optics, etc. Similar to the superconducting chip architecture, they do have advantages and disadvantages. For trapped ions setup, quantum-state engineering or simulation has been shown in chains of 20 trapped ions and 2D crystals of about 300 trapped ions. At present [10], single-qubit initialization can be done with error below $\sim 10^{-3}$, while single-qubit gate error is about 10^{-6} with readout error of $\sim 10^{-4}$. Two-qubit gate error is of $\sim 10^{-3}$. For trapped ions platform, scalability remains the most significant obstacle [17]. Next, nitrogen-vacancy center is one of the numerous point defects present in diamond. Electron spins at the centers can be manipulated at the room temperature with electric, magnetic field, microwave radiation, or a combination. A single nitrogen-vacancy center represents a basic unit of a quantum computer. Multipartite entanglement, long-distance quantum teleportation, quantum error correction, and some basic quantum algorithms have been demonstrated [18] in nitrogen-vacancy-centered diamond. Despite numerous developments in the field, nano-positioning and the creation yield of defects is still a major daunting challenge. On the other hand, a linear-optical quantum computing platform deals with single photons, linear optics elements, photon-counting measurements, etc. Within this platform, even though the experimental control of large entangled states has been demonstrated [19, 20], complete architectures for the platform still need further development and hard bounds on the required performance of photonic devices have to be thoroughly studied in theory [10].

In this thesis, we propose a way to scale up superconducting circuits architecture [21–26]. Similar to cavity quantum electrodynamics (QED), this area of research is commonly referred to as circuit QED. Since every device component in circuit QED is made on chip using standard microfabrication techniques, it offers a number of advantages such as tunability and scalability over the traditional quantum optics setup. Due to the one-dimensional field confinement in the circuit, these systems exhibit large coupling between the artificial atom and the field [25]. Furthermore, we are interested in the fundamental aspects of ultrastrong light–matter coupling (USC) physics, and its use in realizing quantum computing devices. Light–matter interaction in the ultrastrong coupling regime lies at the root of numerous advances in quantum technologies and quantum information tasks [27]. Recent experiments in solid-state physics have reported an unprecedented coupling between a two-level system (qubit) and an optical/microwave cavity [28], reaching the USC [29–40] and deep strong coupling regimes [41], where the light–matter interaction strength is comparable to and larger than the cavity and qubit frequencies, respectively. The USC regime has also been extensively studied in various theoretical contexts [42–52]. Within dipolar approximation, the qubit-cavity system can be described by the quantum Rabi model (QRM) [53, 54], which features a discrete parity symmetry [54, 55]. In this thesis, we will make use of this symmetry from the QRM for quantum computing applications.

In circuit QED context, we regard artificial atoms as qubits. Microwave photons are then used to manipulate and transport quantum information between the artificial atoms, which are stationary at all times. We encode information in the photons (flying qubits) and we intend to manipulate the interaction between the flying qubits and the atoms, based on the on chip tunability of the circuits.

1.1 Structure of the Thesis

The outline of the thesis can be categorized into six parts. In Chap. 2, basics of superconducting circuits are outlined, focusing on the two particular qubits that we use throughout the thesis. In Chap. 3, we introduce USC light–matter interaction and the QRM for the completeness.

No single system is completely isolated from the environment. This is also true for superconducting qubits, especially with their state manipulation, preparation, computation, or even unitary evolution. Thanks to the theory of quantum error correction [56], quantum error correcting codes (QECCs) [57, 58] and the theory of fault-tolerant quantum computation [59], undesired quantum errors can, in principle, be suppressed and corrected in efficient manners. Circuit QED is a prime candidate for implementing large-scale quantum graph codes and other complex QECCs due to their high level of controllability and scalability. Furthermore, the USC regime enables the direct application of ultrafast two-qubit gates [60] between a pair of qubits inside the resonator. With the help of ultrafast gate, we realize quantum error correcting codes in Chap. 4.

Analogous to a classical computer processor, a quantum processor inevitably requires memory cells/devices [61–63] to store arbitrary quantum states in efficient and faithful manner. In particular, these memory devices are needed for storing and retrieving qubits in a fast timescale between the quantum processor and the memory elements, similar to a classical random-access memory. We would also require the ability to store information for a short time with fast storage and readout responses. In Chap 5, we propose a quantum memory implemented on circuit QED with fast storage and retrieval responses, with the memory cell being composed of a two-level system ultrastrongly coupled to the cavity.

It is then followed by a study of the equilibrium and nonequilibrium physics of two qubits mediating through an ultrastrong coupled qubit-cavity system in Chap. 6. By tuning the qubits energy gap and keeping the USC system in its ground state, we will demonstrate a strong two-qubit interaction as well as an enhanced excitation transfer between the two qubits. Finally, we provide some concluding remarks and outlook at some future directions in Chap 7.

References

1. Feynman RP (1982) Simulating physics with computers. Int J Theor Phys 21:467
2. Nielsen MA, Chuang IL (2000) Quantum computation and quantum information. Cambridge University Press
3. Shor PW (1994) Algorithms for quantum computation: discrete logarithms and factoring. In: 35th Annual symposium on foundations of computer science, 1994 Proceedings. IEEE, p 124
4. Grover LK (1997) Quantum mechanics helps in searching for a needle in a haystack. Phys Rev Lett 79:325
5. Bennett CH, Brassard G, Crépeau C, Jozsa R, Peres A, Wootters WK (1895) Teleporting an unknown quantum state via dual classical and Einstein-Podolsky-Rosen channels. Phys Rev Lett 70(13):1993
6. Bennett CH (1995) Quantum information and computation. Phys Today 48:24
7. Bennett CH, Brassard G, Popescu S, Schumacher B, Smolin JA, Wootters WK (1996) Purification of noisy entanglement and faithful teleportation via noisy channels. Phys Rev Lett 76(5):722
8. Ekert AK (1991) Quantum cryptography based on Bell's theorem. Phys Rev Lett 67:661
9. Liao S-K, Cai W-Q, Handsteiner J, Liu B, Yin J, Zhang L, Rauch D, Fink M, Ren J-G, Liu W-Y et al (2018) Satellite-relayed intercontinental quantum network. Phys Rev Lett 120(3):030501
10. Acin A, Bloch I, Buhrman H, Calarco T, Eichler C, Eisert J, Esteve D, Gisin N, Glaser SJ, Jelezko F et al (2018) The quantum technologies roadmap: a european community view. New J Phys 20(8):080201
11. Barends R, Kelly J, Megrant A, Veitia A, Sank D, Jeffrey E, White TC, Mutus J, Fowler AG, Campbell B et al (2014) Superconducting quantum circuits at the surface code threshold for fault tolerance. Nature 508(7497):500
12. Kelly J, Barends R, Fowler AG, Megrant A, Jeffrey E, White TC, Sank D, Mutus JY, Campbell B, Chen Y et al (2015) State preservation by repetitive error detection in a superconducting quantum circuit. Nature 519(7541):66
13. Chen Y, Neill C, Roushan P, Leung N, Fang M, Barends R, Kelly J, Campbell B, Chen Z, Chiaro B et al (2014) Qubit architecture with high coherence and fast tunable coupling. Phys Rev Lett 113:220502
14. Chow JM, Gambetta JM, Magesan E, Abraham DW, Cross AW, Johnson BR, Masluk NA, Ryan CA, Smolin JA, Srinivasan SJ et al (2014) Implementing a strand of a scalable fault-tolerant quantum computing fabric. Nat Commun 5:4015
15. Jeffrey E, Sank D, Mutus JY, White TC, Kelly J, Barends R, Chen Y, Chen Z, Chiaro B, Dunsworth A et al (2014) Fast accurate state measurement with superconducting qubits. Phys Rev Lett 112(19):190504

16. Cross AW, Bishop LS, Sheldon S, Nation PD, Gambetta JM (2018) Validating quantum computers using randomized model circuits. arXiv:1811.12926
17. Monroe C, Kim J (2013) Scaling the ion trap quantum processor. Science 339(6124):1164
18. Waldherr G, Wang Y, Zaiser S, Jamali M, Schulte-Herbrüggen T, Abe H, Ohshima T, Isoya J, Du JF, Neumann P et al (2014) Quantum error correction in a solid-state hybrid spin register. Nature 506(7487):204
19. Meany T, Biggerstaff DN, Broome MA, Fedrizzi A, Delanty M, Steel MJ, Gilchrist A, Marshall GD, White AG, Withford MJ (2016) Engineering integrated photonics for heralded quantum gates. Sci Rep 6:25126
20. Yoshikawa J-I, Yokoyama S, Kaji T, Sornphiphatphong C, Shiozawa Y, Makino K, Furusawa A (2016) Generation of one-million-mode continuous-variable cluster state by unlimited time-domain multiplexing. APL Photonics 1(6):060801
21. Martinis JM, Devoret MH, Clarke J (1985) Energy-level quantization in the zero-voltage state of a current-biased josephson junction. Phys Rev Lett 55(15):1543
22. Nakamura Y, Pashkin YA, Tsai JS (1999) Coherent control of macroscopic quantum states in a single-Cooper-pair box. Naure 398(6730):786
23. You JQ, Nori F (2003) Quantum information processing with superconducting qubits in a microwave field. Phys Rev B 68(6):064509
24. Blais A, Huang R-S, Wallraff A, Girvin SM, Schoelkopf RJ (2004) Cavity quantum electrodynamics for superconducting electrical circuits: an architecture for quantum computation. Phys Rev A 69:062320
25. Wallraff A, Schuster DI, Blais A, Frunzio L, Huang R-S, Majer J, Kumar S, Girvin SM, Schoelkopf RJ (2004) Strong coupling of a single photon to a superconducting qubit using circuit quantum electrodynamics. Nature 431(7005):162
26. Chiorescu I, Bertet P, Semba K, Nakamura Y, Harmans CJPM, Mooij JE (2004) Coherent dynamics of a flux qubit coupled to a harmonic oscillator. Nature 431(7005):159
27. Ladd TD, Jelezko F, Laflamme R, Nakamura Y, Monroe C, O'Brien JL (2010) Quantum computers. Nature 464(7285):45
28. Vahala KJ (2003) Optical microcavities. Nature 424(6950):839
29. Anappara AA, De Liberato S, Tredicucci A, Ciuti C, Biasiol G, Sorba L, Beltram F (2009) Signatures of the ultrastrong light-matter coupling regime. Phys Rev B 79:201303
30. Gunter G, Anappara AA, Hees J, Sell A, Biasiol G, Sorba L, De Liberato S, Ciuti C, Tredicucci A, Leitenstorfer A, Huber R (2009) Sub-cycle switch-on of ultrastrong light-matter interaction. Nature 458(7235):178
31. Todorov Y, Andrews AM, Colombelli R, De Liberato S, Ciuti C, Klang P, Strasser G, Sirtori C (2010) Ultrastrong light-matter coupling regime with polariton dots. Phys Rev Lett 105:196402
32. Scalari G, Maissen C, Turčinková D, Hagenmüller D, De Liberato S, Ciuti C, Reichl C, Schuh D, Wegscheider W, Beck M, Faist J (2012) Ultrastrong coupling of the cyclotron transition of a 2D electron gas to a THz Metamaterial. Science 335(6074):1323
33. Forn-Díaz P, Lisenfeld J, Marcos D, García-Ripoll JJ, Solano E, Harmans CJPM, Mooij JE (2010) Observation of the Bloch-Siegert shift in a qubit-oscillator system in the ultrastrong coupling regime. Phys Rev Lett 105:237001
34. Niemczyk T, Deppe F, Huebl H, Menzel EP, Hocke F, Schwarz MJ, Garcia-Ripoll JJ, Zueco D, Hummer T, Solano E, Marx A, Gross R (2010) Circuit quantum electrodynamics in the ultrastrong-coupling regime. Nat Phys 6:772
35. Schwartz T, Hutchison JA, Genet C, Ebbesen TW (2011) Reversible switching of ultrastrong light-molecule coupling. Phys Rev Lett 106(19):196405
36. Zhang X, Zou C-L, Jiang L, Tang HX (2014) Strongly coupled magnons and cavity microwave photons. Phys Rev Lett 113(15):156401
37. George J, Wang S, Chervy T, Canaguier-Durand A, Schaeffer G, Lehn J-M, Hutchison JA, Genet C, Ebbesen TW (2015) Ultra-strong coupling of molecular materials: spectroscopy and dynamics. Faraday Discuss 178:281
38. George J, Chervy T, Shalabney A, Devaux E, Hiura H, Genet C, Ebbesen TW (2016) Multiple rabi splittings under ultrastrong vibrational coupling. Phys Rev Lett 117(15):153601

39. Forn-Díaz P, García-Ripoll JJ, Peropadre B, Orgiazzi J-L, Yurtalan MA, Belyansky R, Wilson CM, Lupascu A (2017) Ultrastrong coupling of a single artificial atom to an electromagnetic continuum in the nonperturbative regime. Nat Phys 13:39

40. Gao W, Li X, Bamba M, Kono J (2018) Continuous transition between weak and ultrastrong coupling through exceptional points in carbon nanotube microcavity exciton-polaritons. Nat Photonics 12:362

41. Yoshihara F, Fuse T, Ashhab S, Kakuyanagi K, Saito S, Semba K (2017) Superconducting qubit-oscillator circuit beyond the ultrastrong-coupling regime. Nat Phys 13:44

42. Ciuti C, Bastard G, Carusotto I (2005) Quantum vacuum properties of the intersubband cavity polariton field. Phys Rev B 72:115303

43. Ciuti C, Carusotto I (2006) Input-output theory of cavities in the ultrastrong coupling regime: the case of time-independent cavity parameters. Phys Rev A 74:033811

44. Devoret MH, Girvin S, Schoelkopf RJ (2007) Circuit-QED: how strong can the coupling between a Josephson junction atom and a transmission line resonator be? Ann Phys 16(10):767

45. Bourassa J, Gambetta JM, Abdumalikov A, Astafiev O, Nakamura Y, Blais A (2009) Ultrastrong coupling regime of cavity QED with phase-biased flux qubits. Phys Rev A 80:032109

46. Hagenmüller D, De Liberato S, Ciuti C (2010) Ultrastrong coupling between a cavity resonator and the cyclotron transition of a two-dimensional electron gas in the case of an integer filling factor. Phys Rev B 81:235303

47. Nataf P, Ciuti C (2010a) Vacuum degeneracy of a circuit QED system in the ultrastrong coupling regime. Phys Rev Lett 104:023601

48. Douçot B, Feigel'man MV, Ioffe LB, Ioselevich AS (2005) Protected qubits and chern-simons theories in Josephson junction arrays. Phys Rev B 71:024505

49. Nataf P, Ciuti C (2011) Protected quantum computation with multiple resonators in ultrastrong coupling circuit QED. Phys Rev Lett 107:190402

50. Felicetti S, Douce T, Romero G, Milman P, Solano E (2015) Parity-dependent state engineering and tomography in the ultrastrong coupling regime. Sci Rep 5:11818

51. Lolli J, Baksic A, Nagy D, Manucharyan VE, Ciuti C (2015) Ancillary qubit spectroscopy of vacua in cavity and circuit quantum electrodynamics. Phys Rev Lett 114:183601

52. Rossatto DZ, Felicetti S, Eneriz H, Rico E, Sanz M, Solano E (2016) Entangling polaritons via dynamical Casimir effect in circuit quantum electrodynamics. Phys Rev B 93:094514

53. Rabi II (1936) On the process of space quantization. Phys Rev 49:324

54. Braak D (2011) Integrability of the Rabi model. Phys Rev Lett 107:100401

55. Casanova J, Romero G, Lizuain I, García-Ripoll JJ, Solano E (2010) Deep strong coupling regime of the Jaynes-Cummings model. Phys Rev Lett 105:263603

56. Knill E, Laflamme R, Viola L (2000) Theory of quantum error correction for general noise. Phys Rev Lett 84(11):2525

57. Shor PW (1995) Scheme for reducing decoherence in quantum computer memory. Phys Rev A 52(4):R2493

58. Steane AM (1996) Error correcting codes in quantum theory. Phys Rev Lett 77:793

59. Gottesman D (1998) Theory of fault-tolerant quantum computation. Phys Rev A 57:127

60. Romero G, Ballester D, Wang YM, Scarani V, Solano E (2012) Ultrafast quantum gates in circuit QED. Phys Rev Lett 108:120501

61. Pritchett EJ, Geller MR (2005) Quantum memory for superconducting qubits. Phys Rev A 72:010301

62. Reim KF, Michelberger P, Lee KC, Nunn J, Langford NK, Walmsley IA (2011) Single-photon-level quantum memory at room temperature. Phys Rev Lett 107(5):053603

63. Saito S, Zhu X, Amsüss R, Matsuzaki Y, Kakuyanagi K, Shimo-Oka T, Mizuochi N, Nemoto K, Munro WJ, Semba K (2013) Towards realizing a quantum memory for a superconducting qubit: storage and retrieval of quantum states. Phys Rev Lett 111(10):107008

Chapter 2
Basics of Superconducting Circuits Architecture

There is no greatness where there is no simplicity, goodness and truth.

—Leo Tolstoy

In analogy to the cavity QED in quantum optics, the field of circuit QED deals with the study of light and matter interaction on the superconducting chip. In the place of natural atoms and optical laser as in cavity QED, we have artificial atoms and microwave photons propagating on the two-dimensional chip. As this thesis focuses on quantum computation with superconducting qubits, we will briefly look into some basics of the superconducting circuits here.

2.1 An LC Oscillator

It is well known that classical electrical circuits can be modeled and analyzed using the Lagrangian equations [1]. Here, we look at the simplest circuit to apply this formalism.

Let us consider a classical electrical circuit with components: a capacitor ($C = 1$ pF) and an inductor ($L = 10$ nH) as shown in Fig. 2.1. The oscillator has the resonant frequency $\omega_0/2\pi = 1/(2\pi\sqrt{LC}) \approx 1.6$ GHz, which corresponds to a wavelength $\lambda_0 = 20$ cm (microwave domain). However, if the circuit is made small, for instance, in micrometer size, dynamics (the microscopic details) of billions of atoms of the circuit are not important anymore when the resonant microwave is applied to the circuit. The important parameters are then the magnetic flux Φ stored inside the inductor and the charge Q present in the capacitor. This is commonly known as the lumped-element circuit approximation in the literature. Flux belonged to a branch circuit element and is formally defined as

© Springer Nature Switzerland AG 2019
T. H. Kyaw, *Towards a Scalable Quantum Computing Platform in the Ultrastrong Coupling Regime*, Springer Theses,
https://doi.org/10.1007/978-3-030-19658-5_2

Fig. 2.1 An LC oscillator, with the flux Φ at the top node and the ground at the bottom

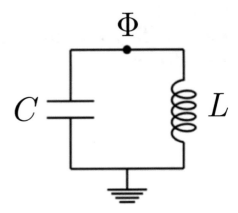

$$\Phi_j(t) = \int_{-\infty}^{t} V_j(t')dt', \tag{2.1}$$

where V_j is is the voltage at branch j. We remark that the flux Φ is related to the phase φ of superconducting condensate's wave function via the following relationship:

$$\varphi = \frac{2e}{\hbar}\Phi \bmod 2\pi, \tag{2.2}$$

which will be derived in Sect. 2.5.2. Whether superconducting or not, the flux can be defined for any circuit. However, the phase φ is usually for superconducting systems. In addition, a branch charge is defined as

$$Q_j(t) = \int_{-\infty}^{t} I_j(t')dt', \tag{2.3}$$

where I_j is the current flow through branch j. In both definition of Φ and Q, we assume that the circuit is at $V_j(-\infty) = I_j(-\infty) = 0$. Suppose the above LC circuit is at ultralow noise environment: $k_B T \ll \hbar\omega_0$, $T = 10$ mK and very high quality factor cavity ($Q > 1000$) (not to be confused with the charge Q). These constraints are normally satisfied for the superconducting circuits made from superconducting materials such as niobium and aluminum, which are then used to fabricate the capacitor and inductor such as shown in Fig. 2.1. Hence, one can promote the variables Φ and Q to be quantum mechanical operators $\hat{\Phi}$ and $\hat{Q}$.[1]

We may calculate the energies stored inside any circuit element at time t

$$E(t) = \int_{-\infty}^{t} V(t')I(t')dt', \tag{2.4}$$

---

[1]Throughout the thesis, it is understood that a variable with a hat on it such as ($\hat{O}$) represents a quantum operator.

by using the the current–voltage relation. One can then write down the Lagrangian of the LC oscillator as

$$\hat{\mathcal{L}} = \frac{1}{2}C\dot{\hat{\Phi}}^2 - \frac{1}{2L}\hat{\Phi}^2. \tag{2.5}$$

Here, $\dot{\Phi}$ means $\frac{\partial \Phi}{\partial t}$ and $\ddot{\Phi}$ means $\frac{\partial^2 \Phi}{\partial t^2}$. The conjugate momentum $\hat{Q}$ to the flux $\hat{\Phi}$ is given by

$$\hat{Q} = \frac{\partial \hat{\mathcal{L}}}{\partial \dot{\hat{\Phi}}} = C\dot{\hat{\Phi}}, \tag{2.6}$$

from which we can derive the Hamiltonian using the Legendre transformation and it is given by

$$\hat{H} = \hat{Q}\dot{\hat{\Phi}} - \hat{\mathcal{L}} = \frac{\hat{Q}^2}{2C} + \frac{\hat{\Phi}^2}{2L}. \tag{2.7}$$

By replacing $\hat{Q} = \hat{p}$, $\hat{\Phi} = \hat{x}$, $C = m$ and $\sqrt{1/LC} = \omega_0$, we then arrive at the Hamiltonian of a quantum harmonic oscillator: $\frac{\hat{p}^2}{2m} + \frac{1}{2}m\omega^2\hat{x}^2$, with the commutator

$$[\hat{\Phi}, \hat{Q}] = i\hbar. \tag{2.8}$$

In addition, the operators $\hat{\Phi}$ and $\hat{Q}$ can be written in terms of the creation and annihilation operators as

$$\hat{\Phi} = \sqrt{\frac{\hbar}{2C\omega_0}}(\hat{a} + \hat{a}^\dagger), \tag{2.9}$$

$$\hat{Q} = -i\sqrt{\frac{\hbar C\omega_0}{2}}(\hat{a} - \hat{a}^\dagger), \tag{2.10}$$

which we note that the following commutation relation holds $[\hat{a}, \hat{a}^\dagger] = 1$. Then, we can rewrite the system Hamiltonian as

$$\hat{H} = \hbar\omega_0\left(\hat{a}^\dagger\hat{a} + \frac{1}{2}\right). \tag{2.11}$$

Hence, we arrive at the familiar quantum harmonic oscillator in terms of the ladder operators, from the classical electrical circuit Lagrangian, after promoting the circuit variables to quantum operators. Of course, it is easy to derive the Hamiltonian for a simple circuit like Fig. 2.1. The complication comes when there are many circuit elements involved. Thus, we need a pedagogical recipe to arrive at quantum Hamiltonian given a circuit, which is the target of the next section.

2.2 Circuit Quantization

Suppose we are given some arbitrary circuit as shown in Fig. 2.2a. There are four steps involved to derive the circuit Hamiltonian with the help of the spanning tree method [2].

1. Choose a preferred node as the ground node from the example circuit, Fig. 2.2a. By doing so, the remaining nodes become the active nodes. See Fig. 2.2b.
2. Starting from the ground node, span around the entire circuit with the constraint that the path cannot form a closed loop (Fig. 2.2c). There, we have chosen a particular spanning tree with the set of tree (T) branches, denoted by red-colored lines. There are also other choices of spanning tree and it does not matter which spanning tree we choose. The set of closure (C) branches, which are not part of the tree, is associated with an irreducible loop. And, a branch flux can be obtained from the associated node fluxes such that

$$\phi_{b \in T} = \Phi_i - \Phi_j, \text{ or} \tag{2.12}$$

$$\phi_{b \in C} = \Phi_i - \Phi_j + \tilde{\Phi}_b, \tag{2.13}$$

where the end nodes i and j belong to the same branch $b \in \{T, C\}$. In addition, $\tilde{\Phi}_b$ is the magnetic flux formed by the inductive elements belonging to the irreducible loop associated with the closure branch.

3. Once chosen the spanning tree, label the branch and node variables as in Fig. 2.2d. The node variables are expressed in terms of branch variables as such

$$\Phi_i = \sum_j \eta_{ij} \phi_j, \tag{2.14}$$

where $\eta_{ij} = \{+1, -1, 0\}$, depending on whether the path connecting the ground to node i follows j (see the arrows) in the same direction ($\eta = +1$), the opposing direction ($\eta = -1$) or does not follow j ($\eta = 0$). For example, $\Phi_6 = \phi_d + \phi_e + \phi_g$, $\Phi_1 = \phi_a$, and $\phi_g = \Phi_6 - \Phi_5 +$ constant.

4. Depending on the number of active nodes present in the circuit, we obtain the same number of Euler–Lagrange equations of motion, with which we can then use to convert to a Lagrangian. After the Legendre transformation, we obtain the Hamiltonian.

2.2.1 An Example

To apply the recipe prescribed above, we look at a simple example shown in Fig. 2.3 [2]. The spanning tree procedure gives us two equations of motion, since there are two active node fluxes Φ_a and Φ_b. They are

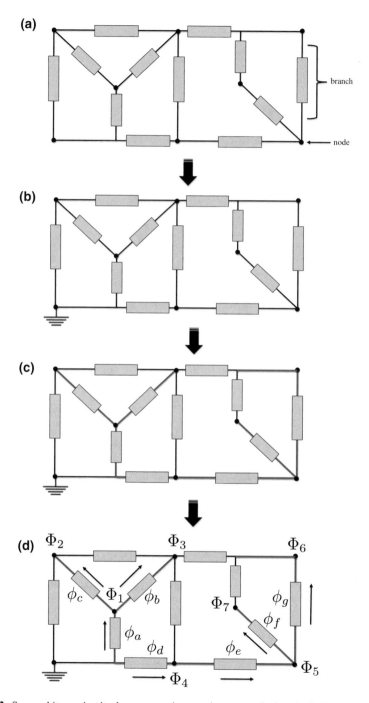

Fig. 2.2 Some arbitrary circuit where we use the spanning tree method to obtain the circuit Hamiltonian. The sequence is from (**a**) to (**d**). A choice of spanning tree is denoted by red-colored lines in (**d**). The arrows denote the current flow in arbitrary choice. Refer to the main text for detailed description

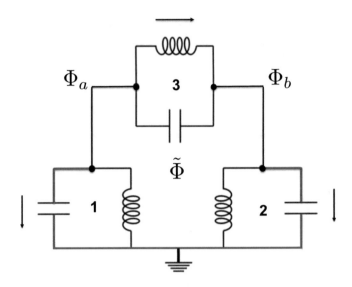

Fig. 2.3 Three LC oscillators are linked up together to form a circuit. We have two active nodes Φ_a, Φ_b and one ground node. The constant $\tilde{\Phi}$ is the net magnetic flux from the three inductive circuit elements L_1, L_2, L_3. A choice of spanning tree is denoted by red-colored lines

$$C_1\ddot{\Phi}_a + \frac{\Phi_a}{L_1} + C_3\left(\ddot{\Phi}_a - \ddot{\Phi}_b\right) + \left(\frac{\Phi_a - \Phi_b + \tilde{\Phi}}{L_3}\right) = 0, \qquad (2.15)$$

$$C_2\ddot{\Phi}_b + \frac{\Phi_b}{L_2} - C_3\left(\ddot{\Phi}_a - \ddot{\Phi}_b\right) - \left(\frac{\Phi_a - \Phi_b + \tilde{\Phi}}{L_3}\right) = 0. \qquad (2.16)$$

These two equations are nothing but the Kirchhoff's Current Law. Due to the conservation of current in the circuit, the system satisfies the Euler–Lagrange's equation:

$$\frac{d}{dt}\left(\frac{\partial \mathcal{L}}{\partial \dot{q}_i}\right) - \frac{\partial \mathcal{L}}{\partial q_i} = 0. \qquad (2.17)$$

For the case above, we have $q_i = \Phi_i$. When deriving Eqs. (2.15) and (2.16), we have made use of the current–voltage relationships for the inductor and the capacitor:

$$V_L(t) = L\frac{dI_L}{dt}, \qquad (2.18)$$

$$I_C(t) = C\frac{dV_C}{dt}, \qquad (2.19)$$

plus the flux definition, Eq. (2.1). Hence, we arrive at the system Lagrangian as

$$\mathcal{L} = \frac{C_1 \dot{\Phi}_a^2}{2} + \frac{C_2 \dot{\Phi}_b^2}{2} + \frac{C_3 (\dot{\Phi}_a - \dot{\Phi}_b)^2}{2} - \left[\frac{\Phi_a^2}{2L_1} + \frac{\Phi_b^2}{2L_2} + \frac{(\Phi_a - \Phi_b + \tilde{\Phi})^2}{2L_3} \right].$$
(2.20)

The Hamiltonian is obtained via the Legendre transformation:

$$H = \sum_{i=a,b} Q_i \dot{\Phi}_i - \mathcal{L},$$
(2.21)

with the conjugate momenta

$$Q_a = \frac{\partial \mathcal{L}}{\partial \dot{\Phi}_a} = C_1 \dot{\Phi}_a - C_3 (\dot{\Phi}_b - \dot{\Phi}_a),$$
(2.22)

$$Q_b = \frac{\partial \mathcal{L}}{\partial \dot{\Phi}_b} = C_2 \dot{\Phi}_b + C_3 (\dot{\Phi}_b - \dot{\Phi}_a).$$
(2.23)

These two equations can be rewritten in a matrix form such that

$$\begin{pmatrix} Q_a \\ Q_b \end{pmatrix} = \underbrace{\begin{pmatrix} C_1 + C_3 & -C_3 \\ -C_3 & C_2 + C_3 \end{pmatrix}}_{M} \begin{pmatrix} \dot{\Phi}_a \\ \dot{\Phi}_b \end{pmatrix},$$

from which we intend to find out $\vec{\dot{\Phi}} = M^{-1} \vec{Q}$. By substituting all the ingredients back to the Hamiltonian, we have

$$H = \frac{1}{C_1 C_2 + C_1 C_3 + C_2 C_3} \left[\frac{(C_2 + C_3) Q_a^2}{2} + \frac{(C_1 + C_3) Q_b^2}{2} \right.$$
(2.24)

$$\left. + \frac{C_3 (Q_b - Q_a)^2}{2} \right] + \left[\frac{\Phi_a^2}{2L_1} + \frac{\Phi_b^2}{2L_2} + \frac{(\Phi_b - \Phi_a + \tilde{\Phi})^2}{2L_3} \right].$$

In this way, a circuit Hamiltonian can be derived. So far, all the variables are classical ones. They can be promoted to quantum operators $\hat{\Phi}$ and $\hat{Q}$, when the system we consider satisfies certain constraints listed in the previous section.

2.3 Transmission Line

In a typical circuit QED experiment such as Fig. 2.4, microwave pulses are guided through a transmission line which is also referred to as the coplanar waveguide (CPW). It is made of a superconducting wire evaporated on an insulating substrate, placed in between superconducting ground planes close to it on the same surface,

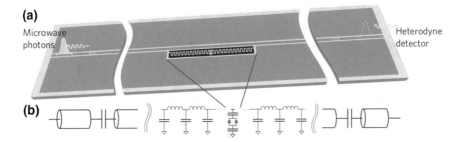

Fig. 2.4 **a** Standard fabrication techniques are used for patterning the simplest circuit QED elements onto a chip: transmission line resonators, superconducting qubits, and coupling capacitors. **b** Capacitive coupling between a qubit and a resonator produces the simplest model for a single-lattice site. (Reproduced with permission from Ref. [3])

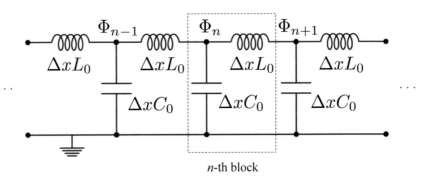

Fig. 2.5 The equivalent circuit of an infinite transmission line with the inductance and capacitance per unit length L_0 and C_0. We have used the node fluxes as the generalized coordinate rather than branch fluxes. The dashed rectangular box represents one block ,which will then repeat itself from the left to right. We can then write its Lagrangian for one block and sum over all such building blocks

as shown in the figure. By coupling many LC oscillators together, a transmission line (part of Fig. 2.4b) is modeled [2] by an infinite chain of electrical circuit shown in Fig. 2.5, where L_0 and C_0 are the inductance and capacitance per unit length of the transmission line. By discretizing the transmission line piecewise in truncated distance Δx and representing in terms of the generalized coordinate, we obtain its Lagrangian as follow.

$$\hat{\mathcal{L}} = \sum_n \left[\frac{1}{2} C_0 \Delta x \dot{\hat{\Phi}}_n^2 - \frac{1}{2} \frac{(\hat{\Phi}_n - \hat{\Phi}_{n-1})^2}{L_0 \Delta x} \right]. \tag{2.25}$$

The charge at the nth node is given by the conjugate momentum

$$\hat{Q}_n = \frac{\partial \hat{\mathcal{L}}}{\partial \dot{\hat{\Phi}}_n} = \Delta x C_0 \dot{\hat{\Phi}}_n. \tag{2.26}$$

By following the recipe in the circuit quantization, Sect. 2.2, we obtain the Hamiltonian

$$\hat{H} = \sum_n \left[\frac{1}{2} \frac{\hat{Q}_n^2}{\Delta x C_0} + \frac{1}{2} \frac{(\hat{\Phi}_n - \hat{\Phi}_{n-1})^2}{L_0 \Delta x} \right], \tag{2.27}$$

which, in the continuum, can be written as

$$\hat{H} = \frac{1}{2} \int_{-\infty}^{\infty} \left[\frac{\hat{Q}(x,t)^2}{C_0} + \frac{1}{L_0} \left(\frac{\partial \hat{\Phi}(x,t)}{\partial x} \right)^2 \right] dx, \tag{2.28}$$

where $\hat{Q}(x,t)$ and $\hat{\Phi}(x,t)$ are the position and time-dependent quantum field operators, with the equal time commutation relations

$$[\hat{Q}(x,t), \hat{Q}(x',t)] = [\hat{\Phi}(x,t), \hat{\Phi}(x',t)] = 0, \tag{2.29}$$
$$[\hat{\Phi}(x,t), \hat{Q}(x',t)] = i\hbar\delta(x-x'). \tag{2.30}$$

Here, $\delta(x-x')$ is the one-dimensional Dirac delta function. We note that the flux field $\hat{\Phi}(x,t)$ satisfies the massless Klein–Gordon equation, which can be expanded in terms of the wave vector k. Using such expression, the Hamiltonian becomes [4]

$$\hat{H} = \int_{-\infty}^{\infty} dk \ \hbar\omega_k \left(\hat{a}_k^\dagger \hat{a}_k + \frac{1}{2} [\hat{a}_k, \hat{a}_k^\dagger] \right), \tag{2.31}$$

with $\omega_k = v|k|$, where $v = 1/\sqrt{L_0 C_0}$ is the propagation velocity. In the above equation, we have the commutator $[\hat{a}_k, \hat{a}_{k'}^\dagger] = \delta(k-k')$.

In experiments, artificial atom couples to one, two (Fig. 2.4), or more transmission line/s. This is also the only mean of communication between experimentalists and the artificial atom. Although transmission line provides key access to the qubit/s, it behaves like an environment, since it can, in principle, accommodate continuum of modes. One such modeling is shown in Appendix B.

2.4 Resonators

By truncating one infinitely long transmission line into a finite-dimensional size, we obtain a resonator, which supports discrete number of standing modes. This concept is analogous to the standing sound waves in a pipe organ, where one can have a pipe with both ends open or a pipe with one end open and another end closed. In the circuit QED language, we can terminate the transmission line either with both

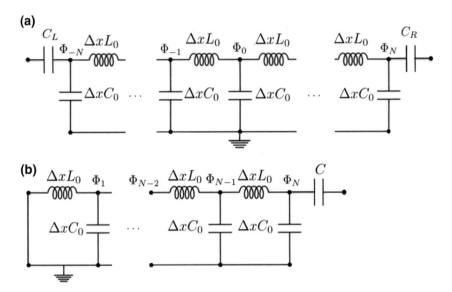

Fig. 2.6 Circuit diagrams for **a** a $\lambda/2$ resonator and **b** a $\lambda/4$ resonator

ends open (each end is connected to a capacitor) or one open end and the opposing end shorted to the ground. The former corresponds to the $\lambda/2$ resonator (Fig. 2.6a) and the latter is the $\lambda/4$ resonator (Fig. 2.6b). Unlike the continuum mode available in the case of a transmission line, we now have restricted discrete modes from the confining boundary conditions, and the Hamiltonian of a resonating cavity is

$$\hat{H} = \sum_n \hbar \omega_n \left(\hat{a}^\dagger(\omega_n)\hat{a}(\omega_n) + \frac{1}{2} \right). \tag{2.32}$$

Here, we have $\omega_n = n\pi v/d$ for $\lambda/2$ resonators, $\omega_n = (n - 1/2)\pi v/d$ for $\lambda/4$ resonators with n being a positive integer number, $v = 1/\sqrt{L_0 C_0}$ the velocity of photons and d the length of the resonator. Most often, we are interested in the fundamental mode $n = 1$, and thus the Hamiltonian becomes that of an LC oscillator, Eq. (2.11).

Since harmonic potential from the LC oscillator gives equidistant energy spectrum, one cannot address individual energy transitions. In order to achieve an artificial two-level system, nonlinearity is needed in the circuit, ensuring anharmonicity in the spectrum. In superconducting circuits, this is accomplished by Josephson junctions.

2.5 Josephson Junction

Dynamics of Josephson junctions and circuits is a very interesting subject matter on its own, and hence interested reader is advised to read the beautiful book by Konstantin Likharev, Ref. [5], for the full glory. In this section, we will briefly go through some interesting phenomena and properties of Josephson junction.

2.5.1 Coherent Phenomena in Superconductivity

In materials, current carriers, either electrons or holes, obey the laws of quantum mechanics. In the standard weakly interacting particles approximation with negligible spin effects, dynamics of each carrier is governed by the Schrödinger equation:

$$i\hbar\dot{\psi} = \hat{H}\psi, \tag{2.33}$$

where ψ is the complex wavefunction of a charge carrier

$$\psi = |\psi(\vec{r}, t)| \exp\{i\varphi(\vec{r}, t)\}. \tag{2.34}$$

$\hat{H}$ is a Hamiltonian and $\hbar$ is the Planck's constant.[2] We know from the quantum physics that $|\psi|^2$ is proportional to the particles density. In the stationary case, the number of particles is fixed $|\psi| \to$ constant, and $\hat{H} \to E$, where E is the energy corresponding to the particular wavefunction ψ. Hence, directly from Eq. (2.33), we arrive at

$$\hbar\dot{\varphi} = -E. \tag{2.35}$$

The quantum nature is now encoded in phase φ of the wavefunction.

In non-superconducting/normal materials, or superconducting materials above its critical superconducting temperature, charge carriers are single electrons or holes. These are fermions obeying the Fermi–Dirac statistics. As a consequence, their energies can never be exactly equal, and so as their rates $\dot{\varphi}$ (c.f. Eq. 2.35). In fact, the phases φ are uniformly distributed along the trigonometric circle [5]. All macroscopic quantities are sum over all the microscopic particles. Hence, in normal conductors, the phase φ is averaged out to zero and it is dropped out of any macroscopic quantity.

However, in a superconducting material well below its superconducting temperature, electrons start to condensate into Cooper pairs, which obey the Bose–Einstein statistics, reaching to their lowest energy level. As a result, their rates $\dot{\varphi}$ are all identical. In addition, Cooper pairs are of relatively large size, $\xi_0 \sim 10^{-4}$ cm, which is much larger than the mean spacing between the pairs ($\sim 10^{-7}$ cm) [5]. That means the wavefunctions of the Cooper pairs overlap to some degree, and all the pairs at a given point in a superconductor are "phase-locked". Therefore, it is sufficient to be

---

[2] $\hbar \approx 1.054 \times 10^{-34}$ Joule-second.

represented by a single wavefunction Ψ (also known as the order parameter of the superconductor):

$$\Psi_j = \sqrt{N_j} \exp(i\varphi_j), \tag{2.36}$$

where N_j is the Cooper pairs density present in the superconducting island j (see Fig. 2.7a, where there are two superconducting islands labeled by φ_1 and φ_2). It is obvious now that the phase φ does not drop out while averaging over all the pairs. Thus, the macroscopic variables such as current can depend on it.

When an external electromagnetic radiation impinges upon the superconductor, the radiation will try to change E and thus the superconducting phase φ has to change in a "quantum mechanical way" satisfying: $\hbar\dot{\varphi} = -E$, which comes from the Schrödinger equation. This direct response leads to

1. the Josephson effect,
2. the magnetic flux quantization,
3. the Meissner effect, and
4. the lack of electrical resistance in superconductor.

For the circuit QED, the Josephson effect gives us nonlinearity in the circuit to achieve artificial qubits.

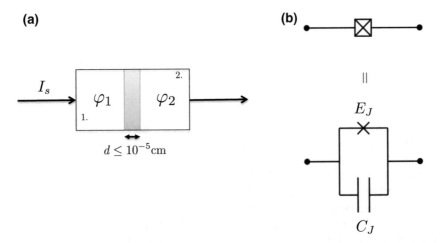

Fig. 2.7 a The cartoon of a Josephson junction: a superconductor–insulator–superconductor (SIS) sandwich is shown. When the electrical contact between the two superconducting islands reaches $d \lesssim 10^{-5}$ cm, the net supercurrent I_s flows through the contact due to phase difference between the two superconducting islands. **b** Circuit diagram of the same junction, which is characterized by its Josephson energy E_J and capacitance C_J. **c** The same Josephson junction in different notation

2.5.2 Josephson Effect

A Josephson junction is an electrical contact of two superconducting islands separated by a thin insulating layer. In other words, it is merely a sandwich of superconductor–insulator–superconductor as shown in Fig. 2.7. Since there are two superconducting islands present in the junction, we have two coherent phases φ_1 and φ_2 for the island 1 and 2. From the Schrödinger equation, we have

$$\hbar\dot{\phi} = E_1 - E_2, \tag{2.37}$$

where $\phi = \varphi_1 - \varphi_2$. The right-hand side of the above equation contains the energy difference of Cooper pairs from the two islands. This difference can exist only if there exists an electrochemical potential between the two islands. Hence,

$$E_1 - E_2 = 2eV, \tag{2.38}$$

which is the energy required to move one Cooper pair from one island to the other. Combining these two equations above, we arrive at the famous "Josephson phase–voltage relation":

$$\dot{\phi} = \frac{2e}{\hbar}V = \frac{2\pi}{\Phi_0}V, \tag{2.39}$$

where $\Phi_0 = \frac{h}{2e}$ is the magnetic flux quantum, the smallest amount of flux a superconducting cylinder or loop can substain.

To derive the Josephson current equation, we make the following intuitive observations adopted from Likharev's book [5]. First, the supercurrent I_s should certainly be dependent on the Cooper-pair density $|\Psi|^2$ in the two superconducting electrodes. Second, for small I_s, it will not change $|\Psi|$, but phases $\varphi_{1,2}$ can change without affecting the physical states of the electrodes. Hence, well-defined current I_s is related to the phase difference $\phi = \varphi_1 - \varphi_2$ alone. Third, 2π phase change should return to the original one, i.e., $I_s(\phi) = I_s(\phi + 2\pi)$. If $\phi = 0$, both electrodes form a single unperturbed superconductor. Therefore, $I_s(\phi) = I_s(0) = I_s(2\pi n) = 0$, where n is the integer multiple. The above three observations imply that I_s has to be of the following form:

$$I_s = I_c \sin(\phi) + \sum_{m=2}^{\infty} I_m \sin(m\phi). \tag{2.40}$$

I_c is the critical current of the Josephson junction. From rigorous theory, the second term, in most cases, can be ignored [5]. We note that Eqs. (2.39) and (2.40) give rise to the AC Josephson effect and the DC Josephson effect, respectively. Even without an external voltage source, there is a current flow due to the difference in superconducting phase.

2.5.3 Energy Storage

Due to the zero-voltage drop across the junction, no energy is dissipated. However, some energy is stored inside the Josephson junction. Consider a process in which the phase changes from $0 \to \phi$. During this process, an external system responsible for the phase change does the following work (from Eq. 2.4) on the supercurrent:

$$W_s = \int_{t_1}^{t_2} I_s V dt = \frac{\hbar I_c}{2e} \int_0^\phi \sin \phi \ d\phi = \frac{\hbar I_c}{2e}(1 - \cos \phi), \qquad (2.41)$$

from which we note that the potential energy has to be in the following form:

$$U_s(\phi) = E_J(1 - \cos \phi), \qquad (2.42)$$

such that $W_s = U_s(\phi) - U_s(0)$, and $E_J = \hbar I_c/(2e)$ is the Josephson energy, a measure of the Cooper pairs ability to tunnel through the junction.

2.5.4 Nonlinear Inductance

The energy storage and conservation in the junction suggests that it can be treated as a nonlinear reactance, an energy-storing two-terminal device. It is also termed as a kinetic inductance, rather than magnetic inductance. The latter stores energy in the form of magnetic field inside an inductor, while the former simply is a result of the presence of supercurrent through the juction.

By differentiating the DC Josephson equation, Eq. (2.40), with time and combining with the AC Josephson equation, Eq. (2.39), we arrive at the Josephson inductance

$$L_J(\phi) = \frac{\Phi_0}{2\pi I_c \cos \phi} = \left(\frac{\Phi_0}{2\pi}\right)^2 \frac{1}{E_J \cos \phi}, \qquad (2.43)$$

after comparing the resultant equation with the Faraday equation: $V_L = L\frac{\partial I}{\partial t}$. We note that the Josephson inductance is not constant, but depends on the phase ϕ across the junction. The most unusual thing is that it can take negative values at the intervals $\pi/2 + 2\pi n < \phi < 3\pi/2 + 2\pi n$.

2.5.5 Josephson Oscillations

Let us consider a case when a nonzero DC voltage V is applied across the Josephson junction. From the Josephson phase–voltage relation, we have

$$\phi = \omega_J t + \text{const.} \tag{2.44}$$

with $\omega_J = (\frac{2e}{\hbar})V = (\frac{2\pi}{\Phi_0})V$. Hence, it corresponds to $I_s = I_c \sin(\omega_J t + \text{const.})$. In a normal inductor, current would just increase gradually with the increase in the voltage V. However, in the case of Josephson junction, the supercurrent oscillates with the increase in applied voltage. The frequency-to-voltage ratio is very high

$$\frac{f_J}{V} = \frac{\omega_J}{2\pi V} = \frac{2e}{\hbar} = \Phi_0^{-1} \approx 483 \, \text{MHz}/\mu\text{V}, \tag{2.45}$$

from which one can estimate $f_J \sim 10^9$–10^{13} Hz at typical voltages of 10^{-6}–10^{-2} V.

2.5.6 Other Current Components

Besides the supercurrent I_s, some other currents can be present too when the operating temperature at which the junction operates is not low, which is normally not the case for the superconducting circuit quantum computing platform. They are displacement current from capacitance, due to the SIS sandwich configuration and the fluctuating current. Therefore, the circuit diagram of the Josephson junction shown in Fig. 2.7b is not valid any more. But, this is beyond the scope of this thesis and the reader is again advised to consult Ref. [5].

Throughout this thesis, we are interested in the low-temperature physics of the Josephson junction with the Lagrangian

$$\hat{\mathcal{L}} = \frac{1}{2} C_J \dot{\hat{\Phi}}^2 + E_J \cos\left(2\pi \frac{\hat{\Phi}}{\Phi_0}\right). \tag{2.46}$$

The potential energy term comes from Eq. (2.42), and the kinetic energy contribution is due to the presence of the capacitor (see Fig. 2.7b).

2.6 DC-SQUID

Additional on-chip tunability is achieved if we take two Josephson junctions placing in a loop as shown in Fig. 2.8, which is basically a DC-superconducting quantum interference device (SQUID). We can then make use of the previous circuit quantization technique to obtain the circuit Lagrangian. However, we need one extra ingredient to get the job done, due to the presence of external magnetic flux. It is commonly known as the fluxoid quantization rule. Experimentally proved in the 1960s [6, 7] due to the condition that the order parameter Ψ is a single-valued [8, 9], in general, the quantization rule suggests that

Fig. 2.8 Circuit diagram of
a DC-SQUID, composed of
a superconducting loop
interrupted by two Josephson
junctions. One can use the
external magnetic flux Φ_{ex} to
tune the effective Josephson
energy of the SQUID. See
the main text for detailed
information

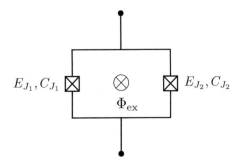

$$\sum_i \Phi_i + \Phi_{\text{ind}} + \Phi_{\text{ex}} = n\Phi_0, \tag{2.47}$$

where Φ_i are node fluxes present in a superconducting loop, Φ_{ind} is the flux associated
with the self-inductance of the SQUID loop, Φ_{ex} is the external applied magnetic
flux threading the loop, and n is the integer multiple. For the DC-SQUID shown
in Fig. 2.8, we have $\Phi_1 - \Phi_2 + \Phi_{\text{ex}} = n\Phi_0$, in the limit of vanishing geometric
inductance of the loop, i.e., $\Phi_{\text{ind}} \to 0$. Here, Φ_1 and Φ_2 are node fluxes associated
to the two junctions. We have neglected Φ_{ind} since it is, in practice, much smaller
as compared to the total junctions inductance. The relation Eq. (2.47) simply means
that the magnetic flux threading the loop is an integer multiple of the fundamental
flux quantum. Without loss of generality, we assume the number of flux quantum in
the loop to be zero. Hence, we have $\Phi_1 = \Phi + \frac{1}{2}\Phi_{\text{ex}}$, and $\Phi_2 = \Phi - \frac{1}{2}\Phi_{\text{ex}}$. Then,
Lagrangian of the DC-SQUID is

$$\hat{\mathcal{L}} = \frac{1}{2}C_{J_1}\dot{\hat{\Phi}}_1^2 + \frac{1}{2}C_{J_2}\dot{\hat{\Phi}}_2^2 + E_{J_1}\cos\left(2\pi\frac{\hat{\Phi}_1}{\Phi_0}\right) + E_{J_2}\cos\left(2\pi\frac{\hat{\Phi}_2}{\Phi_0}\right)$$

$$= \frac{1}{2}C_J\dot{\hat{\Phi}}^2 + E_J(\Phi_{\text{ex}})\cos\left(2\pi\frac{\hat{\Phi}}{\Phi_0}\right), \tag{2.48}$$

where we assume that $C_{J_1} = C_{J_2} = C_J/2$ and $E_{J_1} = E_{J_2} = E_J/2$, with $E_J(\Phi_{\text{ex}}) =$
$E_J\cos\left(\frac{\pi\Phi_{\text{ex}}}{\Phi_0}\right)$, while we have also ignored some constant terms. The expression we
have here is exactly the same as what we have before for the single junction Eq.
(2.46), but with the Φ_{ex} tunable Josephson energy E_J, thereby indirectly being able
to tune the Josephson inductance $L_J = \left(\frac{\Phi_0}{2\pi}\right)^2 \frac{1}{E_J\cos\phi}$.

2.7 Single Cooper-Pair Box and Transmon

After having introduced nonlinear Josephson elements, we are now ready to discuss how one obtains artificial two-level systems from introducing Josephson junctions in circuit. Different superconducting qubits [10] have different ratios of E_J/E_C, where E_J is the Josephson energy, which we have seen earlier and E_C is the charging energy, which will be defined later. Depending on the presiding energy, the qubits can be more sensitive to charge noise when $E_J \ll E_C$ or they are more sensitive to flux noise when $E_J \gg E_C$. The simplest artificial qubit is the single Cooper-pair box (CPB) or charge qubit [11–13], whose effective circuit is shown in Fig. 2.9, and it belongs to the former case. We also note that the circuit for the CPB is similar to that of the LC oscillator shown in Fig. 2.1, except we have replaced the usual inductor with nonlinear circuit element, Josephson junction, in the presence of an external gate voltage. The circuit Lagrangian is

$$\hat{\mathcal{L}} = \frac{1}{2} C_J \dot{\hat{\Phi}}_J^2 + \frac{1}{2} C_g (\dot{\hat{\Phi}}_J + V_g)^2 + E_J \cos\left(2\pi \frac{\hat{\Phi}_J}{\Phi_0}\right), \qquad (2.49)$$

and the conjugate momentum to $\hat{\Phi}$ is

$$\hat{Q}_J = \frac{\partial \hat{\mathcal{L}}}{\partial \dot{\hat{\Phi}}_J} = C_J \dot{\hat{\Phi}}_J + C_g (\dot{\hat{\Phi}}_J + V_g) = 2e\hat{n}. \qquad (2.50)$$

In the last equality, we have assigned the number operator multiplying with $2e$ since a Cooper pair has 2 electrons. n, the eigenvalue of $\hat{n}$, is the number of Cooper pairs on the superconducting island in the charge basis such that $\hat{n}|n\rangle = n|n\rangle$. After the Legendre transformation, we arrive at

Fig. 2.9 Circuit diagram of a single Cooper-pair box qubit, having a voltage source V_g biasing the SCB via the coupling capacitor C_g. The two different colored lines represent two different superconducting islands

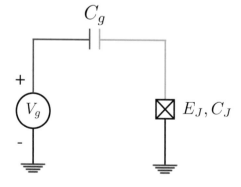

$$\hat{H} = 4E_C(\hat{n} - n_g)^2 - E_J \cos\left(2\pi \frac{\hat{\Phi}_J}{\Phi_0}\right),$$ (2.51)

where we have $E_C = e^2/2(C_g + C_J)$ is the charging energy and $n_g = C_g V_g/2e$ is the dimensionless gate charge or offset charge, representing either the effect of an applied electric field or some microscopic junction asymmetry. The Hamiltonian, Eq. (2.51), can be solved analytically in terms of some special functions. The eigenenergies E_m are

$$E_m(n_g) = E_C a_{2[n_g + k(m,n_g)]}(-E_J/2E_C),$$ (2.52)

where $a_\nu(q)$ denotes Mathieu's characteristic value, and $k(m, n_g)$ is an integer-valued function that orders the eigenvalues [14]. For numerical evaluations, the $a_\nu(q)$ are not easily obtainable. Hence, it is preferable to solve Eq. (2.51) numerically. To go about it, from the commutation relation $[\hat{\Phi}, \hat{Q}] = i\hbar$ (Eq. 2.8), we observe that $\hat{\Phi}_j$ and $\hat{n}$ satisfy the following commutation relation:

$$\left[e^{\widehat{2\pi i\Phi_J/\Phi_0}}, \hat{n}\right] = -e^{\widehat{2\pi i\Phi_J/\Phi_0}}$$ (2.53)

with which one can easily prove the following:

$$e^{\pm\widehat{2\pi i\Phi_J/\Phi_0}}|n\rangle = |n \pm 1\rangle.$$ (2.54)

By rewriting the CPB Hamiltonian in the number of Cooper pairs basis, we then have

$$\hat{H} = \sum_n \left[4E_C(\hat{n} - n_g)^2 |n\rangle\langle n| - \frac{1}{2}E_J\left(|n+1\rangle\langle n| + |n-1\rangle\langle n|\right)\right].$$ (2.55)

Here, we have expanded the cosine term in the original Hamiltonian as sum of two exponentials with the help of Eq. (2.54). One can diagonalize this Hamiltonian numerically and the first three eigen energy levels of the CPB, for three various combinations of E_J/E_C as a function of the gate charge n_g, can be seen in Fig. 2.10. At small ratio E_J/E_C, the energy differences between the levels vary significantly with small change in n_g. That means the qubit there is prone to the charge fluctuation appeared as n_g in the circuit. Even then, one has a well-defined qubit because the two lowest energy levels are far away from all the others, around $n_g = \pm 0.5$, denoted with dashed square boxes in Fig. 2.10.

When the entire circuit is shunted with a large capacitor C_B as shown in Fig. 2.11a, we are at the regime of large E_J/E_C, whose energy spectrum can be seen in Fig. 2.10c. One can clearly see that a good two-level system has disappeared in the spectrum, while the system is now robust against any charge fluctuation. This regime or this kind of artificial atom is commonly known as transmon [14] in the literature and we are going to use it in Chap. 6. We note that the transmon operates in different

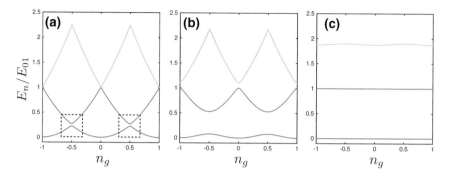

Fig. 2.10 First three eigen energy levels of a single Cooper-pair box in the unit of the energy difference $E_{01} = E_1 - E_0$ are plotted as a function of the gate charge n_g. As the ratio E_J/E_C increases from **a** $E_J/E_C = 0.2$, **b** $E_J/E_C = 2$ to **c** $E_J/E_C = 20$, the charge fluctuation reduces, but the anharmonicity also reduces. In **a**, the two square dashed box indicates the locations of a good artificial two-level system

parameter regime: $E_J/E_C \gg 1$ while CPB is at $E_J/E_C \ll 1$. Therefore, charge is not a good quantum number for transmons just as phase is not a good quantum number for CPBs.

2.8 Flux Qubit

Another widely used superconducting qubit, especially in the ultrastrong coupling (USC) regime, is the flux qubit [15–17]. A brief overview of the flux qubit is summarized here, since we are going to make use of it and its variants throughout this thesis.

The advantage of using flux qubit is that the artificial qubit is robust against charge noise, since the Josephson junctions in the flux qubit design have the ratio $E_J/E_C \gg 1$, but smaller than 100. A typical flux qubit is composed of a small inductance superconducting loop interrupted by three Josephson junctions (see Fig. 2.12a).[3] Among the three junctions, there is one α-junction whose value is in the range of $0.5 < \alpha < 1$. The critical current of that junction is determined by $I_c^{(\alpha)} = \alpha I_c$, where I_c is the critical current of the other two junctions. When the value of external magnetic flux threading the loop is about half the flux quantum, the two lowest energy eigenstates are far away from the rest as seen in Fig. 2.12b. Hence, it can be treated as a qubit. As we will see later, this qubit is characterized by two parameters: energy gap Δ and persistent current I_p within the superconducting loop.

Dynamics of the flux qubit is governed by the following Lagrangian:

---

[3]The main advantages of various multi-Josephson junctions SQUID constructions are that they are much more compact in size and significantly reduce flux noise. In fact, quantum coherence was demonstrated experimentally first in a SQUID with a single E_J-tunable junction [17].

(a)

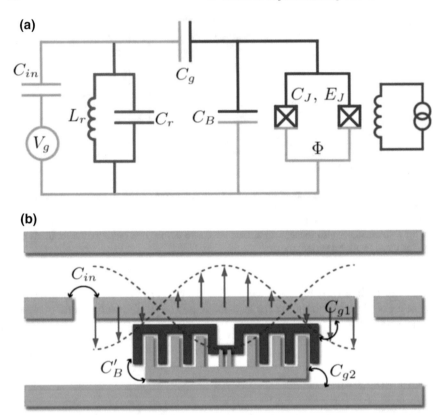

(b)

Fig. 2.11 a Effective circuit diagram of the transmon qubit. The two Josephson junctions (with capacitance and Josephson energy C_J and E_J are shunted by an additional large capacitance C_B, matched by a comparably large gate capacitance C_g. **b** Simplified schematic of the transmon device design (not to scale), which consists of a traditional split Cooper pair box, shunted by a short ($L \sim \lambda/20$) section of twin-lead transmission line, formed by extending the superconducting islands of the qubit. This short section of line can be well approximated as a lumped-element capacitor, leading to the increase in the capacitances C_{g1}, C_{g2}, and C_B' and hence in the effective capacitances C_B and C_g in the circuit. (Reproduced with permission from Ref. [14])

$$\hat{\mathcal{L}} = \sum_i \frac{e^2}{E_{C_i}} \left(\frac{\Phi_0}{2\pi} \right)^2 \dot{\hat{\varphi}}_i^2 + \sum_i E_{J_i}(\cos \hat{\varphi}_i), \qquad (2.56)$$

where we have written the entire equation in terms of the phase difference (branch phase) φ across the junction, rather than Φ, the node flux.[4] In the above equation, we have the Josephson energy $E_{J_i} = \Phi_0 I_{c_i}/(2\pi)$ and the charging energy $E_{C_i} = (2e)^2/(2C_i)$. Since we have a superconducting loop with applied magnetic flux Φ_{ex}, the fluxoid quantization, Eq. (2.47), gives rise to the following phase quantization:

---

[4]We recall that the node phase and node flux are related by the relation $\varphi = \frac{2e}{\hbar} \Phi$ mod 2π.

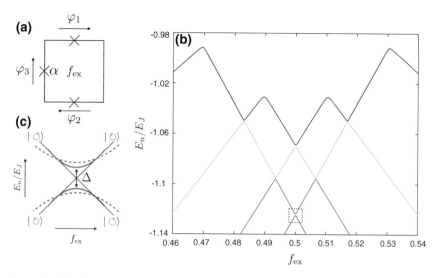

Fig. 2.12 **a** The flux qubit loop interrupted by three Josephson junctions. **b** The four lowest energy eigenstates of the flux qubit are plotted as a function of frustration f_{ex} for $\alpha = 0.75$ and $E_J/E_C = 32$. The dashed square box focuses on the two lowest energy levels which are well separated from the others. **c** The ground state (blue) and the first excited state (red) of the qubit near $f_{ex} = 0.5$. The smaller the α junction becomes, the bigger the energy gap Δ (two dashed curves), and the smaller the persistent current I_p, indicated by slope of the two thin lines

$$\sum_i \varphi_i + 2\pi f_{ex} = 2\pi n, \tag{2.57}$$

where $f_{ex} = \Phi_{ex}/\Phi_0$ is the magnetic frustration. Arriving this phase quantization, we have ignored the self-inductance of the loop because $L_{loop} < \sum_i L_{J_i} = \sum_i \Phi_0/(2\pi I_{c_i})$. From this constraint, we know phase across the α-junction is related to the others via the relation $\varphi_3 = -\varphi_1 - \varphi_2 - 2\pi f_{ex}$. With no loss of generality, $n = 0$ is assumed. Since φ_3 is not an independent variable, it is intuitive to write down the flux qubit potential energy in terms of the sum and difference of the other two phases $\varphi_+ = (\varphi_1 + \varphi_2)/2$ and $\varphi_- = (\varphi_1 - \varphi_2)/2$. Near $f_{ex} = 1/2$, we have

$$U/E_J = 2 + \alpha - 2\cos(\varphi_+)\cos(\varphi_-) + \alpha\cos(\delta f_{ex} + 2\varphi_+), \tag{2.58}$$

where $\delta f_{ex} = f_{ex} - 0.5$ and $|\delta f_{ex}| = 0.5$ corresponds to the maximum frustration. We have plotted the potential energy landscape as a function of the sum and difference of the phases in Fig. 2.13a–c. We note that U is periodic with the period 2π from the figures. From (a to b), we have the frustration $f_{ex} = 0.5$ for the flux qubit with $\alpha = 0.75$ and 0.9, respectively. The larger the value of α, the deeper the two potential wells, which are indicated by two dark blue patches surrounded by the white box. This is obvious when we look at the potential energy cross section in (d). Our qubit has two stable states at the bottom of these energy wells at the

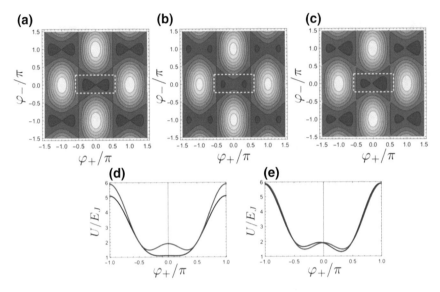

Fig. 2.13 The potential energy of the flux qubit is shown as the functions of the sum and difference of the two phases: φ_+ and φ_-. As seen from (**a–c**), the potential energy is periodic with period 2π, and the two dark blue patches inside the white rectangular dashed boxes indicate the two potential minima. We have **a** $\alpha = 0.75$ and $f_{ex} = 0.5$, **b** $\alpha = 0.9$ and $f_{ex} = 0.5$, **c** $\alpha = 0.75$ and $f_{ex} = 0.6$, respectively. The two potential wells are symmetric at $f_{ex} = 0.5$ (**a–b**), while they become asymmetric when $f_{ex} \neq 0.5$ in **c**. In **d–e**, the potential energy cross section is shown for $\varphi_- = 0$. **d** When α decreases from 0.95 (red) to 0.55 (black) both at $f_{ex} = 0.5$, it is clear that the potential barrier decreases. **e** With the same $\alpha = 0.95$, the asymmetry appears when f_{ex} changes from 0.5 (red) to 0.7 (blue)

locations $\varphi_1 = \varphi_2 = \pm\varphi_+^*$, which determines the clockwise and anticlockwise rotating persistent currents $I_p = \pm I_c \sin(\varphi_+^*)$ inside the loop. The classical magnetic energy associated with these currents are given by $E_{\circlearrowright,\circlearrowleft} = \mp I_p \Phi_0 (f_{ex} - 0.5)$, which can be seen from the two thin lines in Fig. 2.12c. At close to half a flux quantum, the two energy minima locations are independent of the frustration. A little change away from the half flux quantum will tilt the wells as seen in Fig. 2.13c, e. There, the right well is deeper than the left, preferring the state with energy $E_{\circlearrowright} = -I_p \Phi_0 (f_{ex} - 0.5)$.

Quantum mechanics allow the two supercurrent states in the left and right well to interact via quantum mechanical tunneling. Due to this interaction, we observe an avoided level crossing at $f_{ex} = 0.5$ with an energy gap Δ (see Fig. 2.12c) and the eigenstates are symmetric and anti-symmetric superpositions of the two persistent current states $| \circlearrowright \rangle$ and $| \circlearrowleft \rangle$. This tunneling process is dependent on the barrier shape and the kinetic energy available. As we have already seen in Fig. 2.13d, the size of α junction dictates the barrier height. The barrier can be also raised up with increase in E_J since $U \propto E_J$ (c.f. Eq. 2.58), thereby reducing the tunneling probability. On the other hand, increasing the charging energy E_C has the opposing effect. Combining all these effects, the tunnel coupling scales with $\exp(-\alpha E_J/E_C)$ [18]. When $\alpha \leq 0.5$, the barrier vanishes completely for the three junction flux qubit (seen from the black curve of Fig. 2.13d). When $\alpha \to 1$, the inter-cell[5] barriers decrease. Exactly at $\alpha = 1$, the intra- and inter-cell barriers have the same height, which is strongly undesirable. One is advised to suppress the inter-cell tunneling in order to make sure the qubit is insensitive to charge fluctuations occurred on the superconducting islands coupled via the junctions [15].

As seen in Fig. 2.12c, we can apply two-level approximation near the degeneracy point (sweet spot) $f_{ex} = 0.5$. Then, the qubit Hamiltonian can be written in terms of the Pauli matrices $\hat{\sigma}_z$ and $\hat{\sigma}_x$. Here, the z-axis is naturally fixed by and orthogonal to the plane where the superconducting loop is fabricated. In the persistent current basis $(|\circlearrowleft\rangle, |\circlearrowright\rangle)$, the qubit Hamiltonian becomes

$$\hat{H} = -\frac{1}{2}(\epsilon\hat{\sigma}_z + \Delta\hat{\sigma}_x), \tag{2.59}$$

where $\epsilon = 2E_\circlearrowleft = 2I_p(f_{ex} - \frac{1}{2})\Phi_0$. Diagonalizing the Hamiltonian gives the superposed states $|0\rangle$ and $|1\rangle$ (the computational basis) with energy $E_{0/1} = \mp\frac{1}{2}\sqrt{\epsilon^2 + \Delta^2}$ and the energy difference $E_{01} = E_1 - E_0 = \delta = \sqrt{\epsilon^2 + \Delta^2}$. The normalized eigenvectors in the flux basis $(|\circlearrowleft\rangle, |\circlearrowright\rangle)$ are found to be

$$|0\rangle = \frac{1}{\sqrt{\delta(2\delta - 2\epsilon)}}\begin{pmatrix} \Delta \\ \delta - \epsilon \end{pmatrix}, \tag{2.60}$$

$$|1\rangle = \frac{1}{\sqrt{\delta(2\delta - 2\epsilon)}}\begin{pmatrix} \delta - \epsilon \\ -\Delta \end{pmatrix}. \tag{2.61}$$

We have so far discussed a short overview of the charge qubit, transmon, and flux qubit. However, there are still many other superconducting qubits [10] in the literature. We are not able to review them here, which is beyond this thesis scope. At the time of this thesis writing, besides the three qubits mentioned above, there are phase qubit [19–21], quantronium [22], fluxonium [23], Xmon [24], and Gmon [25], etc. One can see their individual performances in summary Fig. 2.14. For the latest development of very good quality superconducting qubits besides Google and IBM, one is advised to refer to Ref. [26], which shows comparable device performance to Google superconducting qubits.

---

[5]The potential minima, enclosed by the white dashed box, repeat in period 2π. Each box is considered one cell.

2.9 Mendeleev-Like Table for Superconducting Qubits

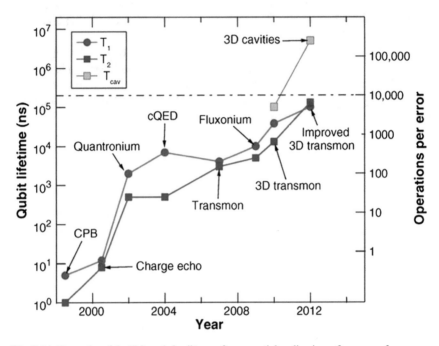

Fig. 2.14 Examples of the "Moore's law" type of exponential scaling in performance of superconducting qubits during recent years. Improvement of coherence times for the "typical best" results associated with the first versions of major design changes. The blue, red, and green symbols refer to qubit relaxation, qubit decoherence, and cavity lifetimes, respectively. Innovations were introduced to avoid the dominant decoherence channel found in earlier generations. So far an ultimate limit on coherence seems not to have been encountered. For comparison, superconducting cavity lifetimes are given for a 3D transmon and separate 3D cavities [27]. (Reproduced with permission from Ref. [28])

References

1. Wells DA (1938) Application of the Lagrangian equations to electrical circuits. J Appl Phys 9(5):312
2. Devoret MH (1995) Quantum fluctuations in electrical circuits. In: Session LXIII, Les Houches, vol 7, no 8
3. Houck AA, Türeci HE, Koch J (2012) On-chip quantum simulation with superconducting circuits. Nat Phys 8(4):292
4. Peskin ME, Schroeder DV (1995) An introduction to quantum field theory. Westview Press Incorporated
5. Likharev KK (1986) Dynamics of Josephson junctions and circuits. Gordon and Breach Science Publishers

6. Doll R, Näbauer M (1961) Experimental proof of magnetic flux quantization in a superconducting ring. Phys Rev Lett 7(2):51
7. Deaver BS Jr, Fairbank WM (1961) Experimental evidence for quantized flux in superconducting cylinders. Phys Rev Lett 7(2):43
8. London F (1954) Superfluids. Wiley
9. Onsager L (1961) Magnetic flux through a superconducting ring. Phys Rev Lett 7(2):50
10. Clarke J, Wilhelm FK (2008) Superconducting quantum bits. Nature 453(7198):1031
11. Bouchiat V, Vion D, Joyez Ph, Esteve D, Devoret MH (1998) Quantum coherence with a single Cooper pair. Phys Scr 1998(T76):165
12. Makhlin Y, Scöhn G, Shnirman A (1999) Josephson-junction qubits with controlled couplings. Nature 398:305
13. Nakamura Y, Pashkin YA, Tsai JS (1999) Coherent control of macroscopic quantum states in a single-Cooper-pair box. Naure 398(6730):786
14. Koch J, Yu TM, Gambetta J, Houck AA, Schuster DI, Majer J, Blais A, Devoret MH, Girvin SM, Schoelkopf RJ (2007) Charge-insensitive qubit design derived from the Cooper pair box. Phys Rev A 76:042319
15. Orlando TP, Mooij JE, Tian L, van der Wal CH, Levitov LS, Lloyd S, Mazo JJ (1999) Superconducting persistent-current qubit. Phys Rev B 60:15398
16. Mooij JE, Orlando TP, Levitov L, Tian L, van der Wal CH, Lloyd S (1999) Josephson persistent-current qubit. Science 285(5430):1036
17. Friedman JR, Patel V, Chen W, Tolpygo SK, Lukens JE (2000) Quantum superposition of distinct macroscopic states. Nature 406(6791):43
18. Paauw FG (2009) Superconducting flux qubits: quantum chains and tunable qubits. PhD thesis, TU Delft, Delft University of Technology
19. Martinis JM, Devoret MH, Clarke J (1985) Energy-level quantization in the zero-voltage state of a current-biased Josephson junction. Phys Rev Lett 55(15):1543
20. Clarke J, Cleland AN, Devoret MH, Esteve D, Martinis JM (1988) Quantum mechanics of a macroscopic variable: the phase difference of a Josephson junction. Science 239(4843):992
21. Martinis JM, Nam S, Aumentado J, Urbina C (2002) Rabi oscillations in a large Josephson-junction qubit. Phys Rev Lett 89(11):117901
22. Vion D, Aassime A, Cottet A, Joyez Pl, Pothier H, Urbina C, Esteve D, Devoret MH (2002) Manipulating the quantum state of an electrical circuit. Science 296(5569):886
23. Manucharyan VE, Koch J, Glazman LI, Devoret MH (2009) Fluxonium: single Cooper-pair circuit free of charge offsets. Science 326(5949):113
24. Barends R, Kelly J, Megrant A, Sank D, Jeffrey E, Chen Y, Yin Y, Chiaro B, Mutus J, Neill C et al (2013) Coherent Josephson qubit suitable for scalable quantum integrated circuits. Phys Rev Lett 111(8):080502
25. Chen Y, Neill C, Roushan P, Leung N, Fang M, Barends R, Kelly J, Campbell B, Chen Z, Chiaro B et al (2014) Qubit architecture with high coherence and fast tunable coupling. Phys Rev Lett 113:220502
26. Gong M, Chen MC, Zheng Y, Wang S, Zha C, Deng H, Yan Z, Rong H, Wu Y, Li S et al (2018) Genuine 12-qubit entanglement on a superconducting quantum processor. arXiv:1811.02292
27. Reagor M, Paik H, Catelani G, Sun L, Axline C, Holland E, Pop IM, Masluk NA, Brecht T, Frunzio L et al (2013) Reaching 10 ms single photon lifetimes for superconducting aluminum cavities. Appl Phys Lett 102(19):192604
28. Devoret MH, Schoelkopf RJ (2013) Superconducting circuits for quantum information: an outlook. Science 339(6124):1169

Chapter 3
Ultrastrong Light–Matter Interaction

The Tao that can be told of, is not the Absolute Tao;
The Names that can be given, are not Absolute Names.

—Lao Tzu

Einstein once wrote [1]: "The wave theory of light which operates with continuous functions in space has been excellently justified for the representation of a purely optical phenomena and it is unlikely ever to be replaced by another theory. One should, however, bear in mind that optical observations refer to time averages and not to instantaneous values and notwithstanding the complete experimental verification of the theory of diffraction, reflection, refraction, dispersion and so on, it is quite conceivable that a theory of light involving the use of continuous functions in space will lead to contradictions of experience, if it is applied to the phenomena of the creation and conversion of light".

As will be seen later in this chapter, an artificial atom strongly coupled to the resonator mode inside a cavity is exactly one such situation. A natural atom strongly coupled to cavity mode is also another example. Furthermore, the atom–cavity system is uniquely pertinent to the efficient single photons generation, a phenomenon where Einstein's "conflict of experience" inevitably dominate. Therefore, the theory of light quantization is needed to describe cavity quantum electrodynamics (cavity QED).

3.1 Cavity Quantum Electrodynamics

Cavity QED explores the light–matter interaction inside an electromagnetic resonator where quantized electromagnetic fields and atomic systems coherently interact. A simple setup is shown in Fig. 3.1, where two-level atomic system remains at rest inside a resonator mode formed by two spherical mirrors. The system dynamics

© Springer Nature Switzerland AG 2019
T. H. Kyaw, *Towards a Scalable Quantum Computing Platform in the Ultrastrong Coupling Regime*, Springer Theses,
https://doi.org/10.1007/978-3-030-19658-5_3

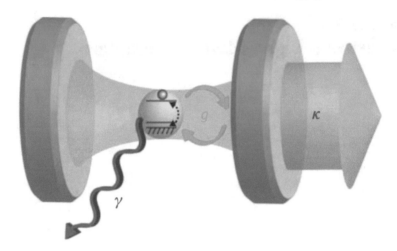

Fig. 3.1 Elements of cavity QED. Shown is a simple schematic of an atom–cavity system depicting the three governing rates (g, κ, γ) in cavity QED. Coherent exchange of excitation between the atom and the cavity field proceeds at rate g, as indicated by the dashed arrow for the atom and the green arrows for the cavity field. (Reproduced with permission from Ref. [2])

is well-studied in literature [3]. One possibility is to quantize the electromagnetic radiation inside the cavity in the Coulomb gauge. Assuming the dipolar interaction between the atom and the radiation, the minimal coupling Hamiltonian is achieved. By limiting that there is one single radiation mode present in the cavity that is resonant with the first two atomic energy level transition, the quantum Rabi model (QRM) [4, 5] is obtained. Typically, in cavity QED experiments, the light–matter interaction is not so strong.[1] Thus, in the interaction picture, the rotating wave approximation (RWA) can be applied and the well-known Jaynes–Cummings (JC) Hamiltonian [6] is obtained.

$$\hat{H}_{\text{JC}} = \frac{\hbar}{2}\omega_q \hat{\sigma}_z + \hbar\omega_{\text{cav}}\hat{a}^\dagger \hat{a} + \hbar g(\vec{r})\left(\hat{a}^\dagger\hat{\sigma}_- + \hat{a}\hat{\sigma}_+\right). \tag{3.1}$$

The Hamiltonian is composed of three parts. The first two corresponds to the atom and field, while the last one represents the atom–photon interaction. $\hat{\sigma}$'s are usual Pauli operators and $\hat{a}^\dagger(\hat{a})$ is creation (annihilation) operator of the single-mode bosonic quantized field inside the cavity. ω_q and ω_{cav} are the resonant frequencies belonged to the two-level atom and the cavity, respectively. The coupling magnitude is given by a function of the atom's position $\vec{r}$ within the standing wave structure of the mode [7]:

$$g(\vec{r}) = \sqrt{\frac{\mu^2 \omega_{\text{cav}}}{2\hbar\epsilon_0 V_M}}\, U(\vec{r}) \equiv g_0 U(\vec{r}), \tag{3.2}$$

---

[1]We will elaborate more on the strong light–matter criterion in Sect. 3.1.1.

where μ is the atomic dipole matrix element, $U(\vec{r})$ is the cavity mode function, such that $V_M = \int |U(\vec{r})|^2 d^3x$, and V_M is the resonant mode volume. The single-photon Rabi frequency $2g_0$ represents the rate at which single quantum of excitation is exchanged between light and matter [7].

Unlike the ideal setup discussed above, there are two common dissipative processes that naturally arise in real atoms and cavities. The first one is the photons leakage through the two mirrors at a rate of 2κ, where κ is the frequency half-width of the resonant mode. It is associated with the quality factor of the cavity, the quantity that specifies how many times a photon can be reflected by the mirrors inside the cavity before it leaks out.

$$Q = \frac{\omega_{cav}}{2\kappa}. \tag{3.3}$$

The second dissipation is due to the spontaneous emission from the atom into other field modes rather than the preferentially coupled cavity mode. Usually, there are two atomic decay rates along the transverse and longitudinal directions of the cavity $(\gamma_\perp, \gamma_\parallel)$. And, they are dependent on where the atom is located inside the cavity $(\vec{r})$. However, this position dependency is commonly negligible for the state-of-the-art optical Fabry–Perot cavities. In this scenario, the transverse decay rate is very well approximated by the atomic free space decay rate: $\gamma_\perp = \gamma = \gamma_\parallel/2$ [7].

3.1.1 Strong Light–Matter Interaction

As seen above, cavity QED depends on three rates (g_0, κ, γ), where we have not mentioned the amount of time an atom resides inside a cavity denoted by T, since we have assumed stationary atoms. This is the case in superconducting circuits. In general, T also an important factor in cavity QED. The strong coupling criterion [8] suggests that the coupling coefficient dominates dissipation:

$$g_0/(\gamma, \kappa, T^{-1}) \gg 1. \tag{3.4}$$

Strong coupling can be corroborated with two dimensionless parameters, namely

1. the saturation photon number and
2. the critical atom number.

The former represents the number of photons such that the optical intensity, for a cavity of a given geometry, is sufficient to saturate the atom:

$$n_0 = \frac{\gamma^2}{2g_0^2}. \tag{3.5}$$

The latter is the number of strongly coupled atoms required to appreciably affect the cavity field:

$$N_0 = \frac{2\kappa\gamma}{g_0^2}. \tag{3.6}$$

In quantum optics, lasers, for example, have $\sqrt{n_0} \sim 10^3 - 10^4$, large critical parameters. Hence, the coherent coupling g_0 tends to be small as processes tend to semiclassical limit. In contrast, the necessary but not sufficient criteria for strong coupling are

$$(n_0, N_0) \ll 1, \tag{3.7}$$

which means single quanta dominate the dynamics of the system in the strong coupling. Observation of the vacuum Rabi splitting is one clear indicator.

When the cavity mode is resonant with the atomic transition frequency ($\omega_{cav} = \omega_q$), exact diagonalization of the JC Hamiltonian, Eq. (3.1), yields a set of eigenstates known as Jaynes–Cummings doublets:

$$|\pm, n\rangle = \frac{1}{\sqrt{2}}(|g, n\rangle \pm |e, n-1\rangle), \tag{3.8}$$

where $|g, e\rangle$ denote the ground and excited states of the two-level system and $|n\rangle$ is n-number of photons inside the cavity, in the Fock state basis. The above dressed states have corresponding eigen energies

$$E_{\pm,n} = n\hbar\omega_{cav} \pm \sqrt{n}\hbar g. \tag{3.9}$$

Experimentally, one can see two-peaked structure (Fig. 3.2a) with maxima at $\omega = \omega_{cav} \pm g$, corresponding to the states $|\pm, 1\rangle$ for a single excitation quantum (transitions indicated by the red lines in Fig. 3.2b), rather than a single empty cavity resonance at ω_{cav} (red dashed lines in (a)). This characteristic spectral feature is known as the vacuum Rabi splitting and serves as a hallmark of strong coupling. It has been observed both in cavity QED [10] and in circuit QED [11].

In the regime where the atom and cavity frequencies are not resonant, i.e., $\delta \equiv \omega_q - \omega_{cav}$, a Schrieffer-Wolf transformation (see Appendix C) can be applied to the JC Hamiltonian if the dispersive condition is met: $g/\delta \ll 1$, up to second order in g^2/δ [12], attaining the AC Stark Hamiltonian:

$$\hat{H}_{AC} = \frac{\hbar}{2}\left[\omega_q + \frac{g^2}{\delta}\right]\hat{\sigma}_z + \hbar\left[\omega_{cav} + \frac{g^2}{\delta}\hat{\sigma}_z\right]\hat{a}^\dagger\hat{a}. \tag{3.10}$$

This light–matter coupling is manifested in the non-radiative energy shifts that atom and field exert on each other. Detecting the cavity field yields a quantum nondemolition measurement of the qubit state. As seen in the above equation, in the second term, the field frequency is shifted with the amount $\left[\omega_{cav} \pm \frac{g^2}{\delta}\right]$ depending on the state of the qubit. This approach is widely used in superconducting circuits quantum computing experiments [13].

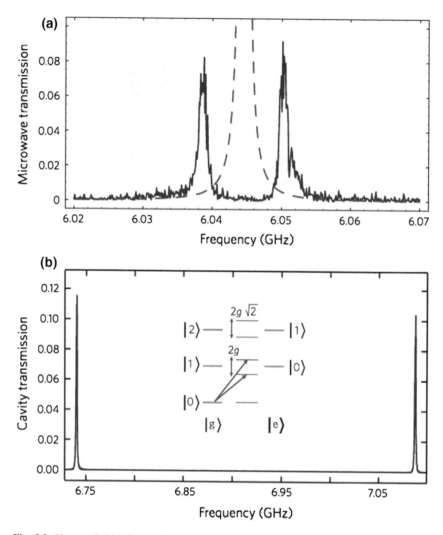

Fig. 3.2 Vacuum Rabi splitting. Observation of strong coupling and the fine-structure limit in a circuit. **a** Measurement of the microwave transmission of a cavity like that in Fig. 3.3b. The appearance of two peaks in the transmission, as a result of vacuum Rabi splitting, indicates strong coupling. Without the qubit, a single transmission peak (dashed line) is observed. With the qubit tuned to match the cavity frequency, the qubit–cavity interaction mixes together the photon and qubit states, and the new eigenstates of the system are coherent superpositions that are symmetric and antisymmetric combinations of atom and photon. The decay rates of these half-atom/half-photon superposition states are the average of the photon and atom decay rate, $(g + \kappa)/2$. Strong coupling is observed by starting with the system in its lowest energy state (with no photons and the atom in the ground state) and measuring the presence of two peaks separated by $2g \sim 12$ MHz about the original cavity resonance. **b** A more recent experimental result, showing a separation of the vacuum Rabi peaks by about $2g/2\pi = 350$ MHz, in which $g/\omega \sim 2.5\%$; the cavity decay rate is $\kappa/2\pi \sim 800$ kHz and the qubit decay rate is $\gamma/2\pi \sim 200$ kHz. This experiment approaches the fine-structure limit for the maximal value of an electric dipole coupling. (Reproduced with permission from Ref. [9])

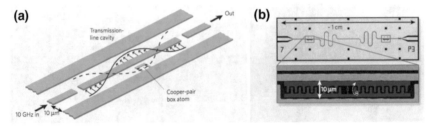

Fig. 3.3 a Schematic representation of the circuit analogue of cavity QED, where a superconducting qubit (green) interacts with the electric fields (pink) in a transmission line (blue), consisting of a central conductor and two ground planes on either side. The cavity is defined by two gaps (the mirrors) separated by about a wavelength. The cavity and qubit are measured by sending microwave signals down the cable on one side of the cavity and collecting the transmitted microwaves on the output side. **b** Micrograph of an actual circuit QED device that achieves the strong coupling limit. It consists of a superconducting niobium transmission line on a sapphire substrate with two qubits (green boxes) on either side. The inset shows one of the superconducting Cooper-pair box charge qubits located at the ends of the cavity where the electric fields are maximal. The qubit has two aluminum "islands" connected by a small Josephson junction. Changing the state of the qubit corresponds to moving a pair of electrons from the bottom to top (shown schematically). (Reproduced with permission from Ref. [9])

3.2 Circuit Implementation of Cavity QED

We now consider the superconducting circuits as shown in Fig. 3.3 to realize the same Jaynes–Cummings physics we have seen with cavity QED system in the previous section. A one-dimensional transmission line resonator consisting of superconducting coplanar waveguide plays the cavity role and a superconducting qubit (transmon) plays the atom role. The Hamiltonian of the combined system [12] (see Chap. 2) is given by

$$\hat{H} = \hbar\omega_r \left(\hat{a}^\dagger \hat{a} + \frac{1}{2} \right) + \frac{\hbar\omega_q}{2} \hat{\sigma}_z - e\frac{C_g}{C_\Sigma} \sqrt{\frac{\hbar\omega_r}{cL}} \left(\hat{a}^\dagger + \hat{a} \right)$$
$$\times \left[1 - 2N_g - \cos(\theta)\hat{\sigma}_z + \sin(\theta)\hat{\sigma}_x \right]. \tag{3.11}$$

Here, $\hat{\sigma}$'s are Pauli matrices in the eigenbasis of $\{| \uparrow\rangle, | \downarrow\rangle\}$, $\theta = \arctan[E_J/4E_C(1 - 2N_g^{dc})]$, the qubit energy is $\omega_q = \sqrt{E_J^2 + [4E_C(1 - 2N_g^{dc})]^2}/\hbar$ and c is the capacitance per unit length of the CPW resonator. At the charge degeneracy point where $N_g^{dc} = C_g V_g^{dc}/2e = 1/2$ and $\theta = \pi/2$, the above Hamiltonian reduces to the JC Hamiltonian, Eq. (3.1), after neglecting rapidly oscillating terms and damping. There, $\omega_q = E_J/\hbar$ and the light–matter coupling term is

$$g = \frac{C_g e}{\hbar C_\Sigma} \sqrt{\frac{\hbar\omega_r}{cL}}. \tag{3.12}$$

Here, L is the length of the resonator seen in Fig. 3.3a. Its start and end points are defined by the two mirrors. In this kind of circuit, the artificial atom is highly polarizable at the charge degeneracy point, having transition dipole moment $d \sim 2 \times 10^4$ atomic units, or more than an order of magnitude larger than even a typical Rydberg atom [12]. We will see in the next section that one can in fact obtain the light–matter interaction which is beyond the reach of quantum optical systems, by engineering superconducting atom inside the resonator, opening up many new possibilities and interesting physics.

3.3 Quantum Rabi Model

As we have seen previously, the quantum Rabi model (QRM) naturally arises from the minimal coupling Hamiltonian

$$\hat{H}_{\mathrm{QRM}} = \frac{\hbar\omega_q}{2}\hat{\sigma}_z + \hbar\omega_r\hat{a}^\dagger\hat{a} + \hbar g\hat{\sigma}_x(\hat{a} + \hat{a}^\dagger). \tag{3.13}$$

It was first introduced by Isidor Rabi in 1936 [4], and an integrability of the model was recently made known by Daniel Braak in 2011 [5]. Unlike the JC model, it includes the counterrotating terms which were neglected in the JC model due to strong light–matter coupling. When the coupling is comparable to the cavity or qubit frequency, i.e., $0.1 \leq g/\omega_r < 1$, the counterrotating terms start to play an important role. As seen from Fig. 3.4, experimentally measured transmission through the circuit with a flux qubit galvanically coupled to the resonator would not be matched with theoretical predictions if the counterrotating terms were neglected. Figure 3.4c–d show clearly the stark deviation from the experimental results with the ones predicted from the JC model (solid lines). We remark that the ultrastrong coupling regime was experimentally realized not only in superconducting circuits [14–17] but also in semiconductor quantum wells–THz metamaterial hybrid [18], carbon nanotubes [19], molecules in optical cavities [20–22] as well as magnons in microwave cavities [23]. Interested reader is advised to consult the recent review article [24] and references therein.

Unlike the JC model where the conserved quantity is the total number of excitations, the QRM has a discrete $\mathbb{Z}_2$ parity symmetry denoted by the parity operator

$$\hat{P} = \hat{\sigma}_z e^{i\pi\hat{a}^\dagger\hat{a}}, \tag{3.14}$$

which has ± 1 eigenvalues. As a result, the total Hilbert space can be split into two infinite chains [25] belonged to two different parity eigenvalues:

$$| \downarrow, 0\rangle \leftrightarrow |\uparrow, 1\rangle \leftrightarrow |\downarrow, 2\rangle \leftrightarrow |\uparrow, 3\rangle \leftrightarrow \cdots (p = -1), \tag{3.15}$$
$$| \uparrow, 0\rangle \leftrightarrow |\downarrow, 1\rangle \leftrightarrow |\uparrow, 2\rangle \leftrightarrow |\downarrow, 3\rangle \leftrightarrow \cdots (p = +1).$$

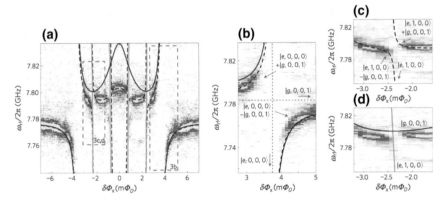

Fig. 3.4 Experimental observation of transitions that violates the excitations number conservation in a flux qubit–resonator circuit. The figures show transmission through the circuit with ω_{rf} being the probe frequency and $\delta\Phi_x$ being the external flux applied to the qubit. **a** Full circuit spectrum near second resonator mode frequency. Dashed lines fitting the data corresponding to the full Hamiltonian, the green vertical lines correspond to the case of no qubit–resonator coupling, while the solid dark blue line comes from the Jaynes–Cummings model prediction; **b** zoom in near avoided qubit–resonator level crossing; **c** avoided level crossings not included in the JC model. It is evident that the presence of the counterrotating terms introduces hybridization between the labeled eigenstates that otherwise would not couple. (Reproduced with permission from Ref. [15])

The QRM can also be written in terms of the parity operator

$$\hat{H}_{\text{QRM}} = \hbar\omega_r \hat{b}^\dagger \hat{b} + \hbar g\left(\hat{b} + \hat{b}^\dagger\right) - \hbar\frac{\omega_q}{2}(-1)^{\hat{b}^\dagger \hat{b}}\hat{P}, \qquad (3.16)$$

where $\hat{b} = \hat{\sigma}_x \hat{a}$.

Recently, superconducting qubit–oscillator circuit beyond the ultrastrong coupling regime has been realized in the experiment [17], by inductively coupling a flux qubit and an LC oscillator via Josephson junctions as shown in Fig. 3.6. A very nice trick, employed in both of the experiments shown in Figs. 3.5 and 3.6, is to insert a very large inductor in the connection between qubit and resonator, thereby effectively increases the coupling and reaches the deep-strong coupling regime [25]. This approach is known as the galvanic coupling, which will be discussed in next chapter.

All the various light–matter coupling regimes from cavity QED to recent circuit QED experiments are summarized in the Table.

	Weak	Strong	Ultrastrong	Deep strong
Coupling	$g < \max\{\gamma, \kappa\}$	$g > \max\{\gamma, \kappa\}$	$g \gtrsim 0.1\omega_{\text{cav/qubit}}$	$g \gtrsim \omega_{\text{cav/qubit}}$
Model	JC model	JC model	QRM	QRM

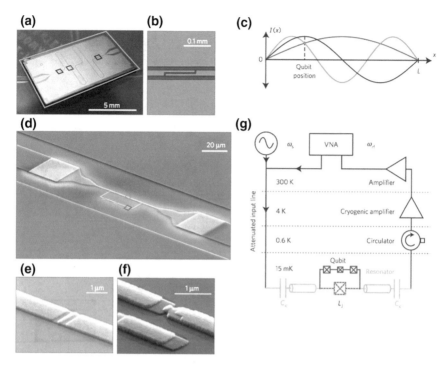

Fig. 3.5 Quantum circuit and experimental set up. **a** Optical image of the superconducting $\lambda/2$ coplanar waveguide resonator (light blue rectangle). Black rectangles: area shown in **b** Red rectangle: area shown in **d**. **b** SEM image of one of the coupling capacitors. **c** Sketch of the current distribution of the first three resonator modes. Their resonance frequencies are $\omega_1/2\pi = 2.782$ GHz ($\lambda/2$, red), $\omega_2/2\pi = 5.357$ GHz (λ, blue) and $\omega_3/2\pi = 7.777$ GHz ($3\lambda/2$, green). The cavity modes ω_n are measured at maximum qubit–cavity detuning ($\Phi_x = 0$). In general, the flux dependence of ω_n is very weak, except for the regions close to $\Phi_x = \pm\Phi_0/2$. **d** SEM image of the galvanically coupled flux qubit. The width in the overlap regions with the center conductor is 20 μm, and that of the constriction is 1 μm. Orange rectangle: area shown in **e**. Green rectangle: area shown in **f**. **e** SEM image of the large Josephson junction. Its Josephson inductance L_J is responsible for approximately 85% of the qubit–resonator coupling. **f** One Josephson junction of the qubit loop. The area of this junction is 14% of the one shown in **e**. **g** Schematic of the measurement set up. The transmission through the cavity at ω_{rf} is measured using a vector network analyser (VNA). A second microwave signal at ω_s is used for two-tone qubit spectroscopy. The input signal is attenuated at various temperature stages and coupled into the resonator (light blue) through the capacitors C_κ. The crossed squares represent Josephson junctions. A circulator isolates the sample from the amplifier noise. (Reproduced with permission from Ref. [15])

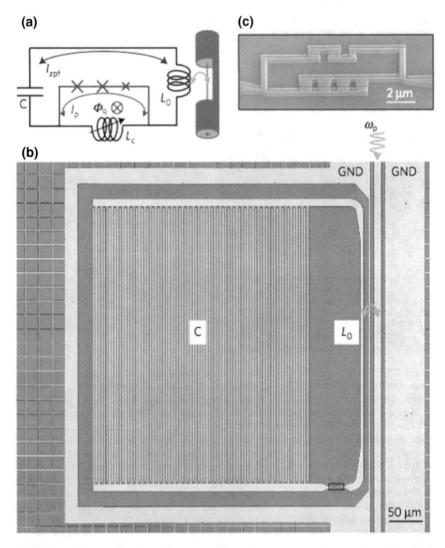

Fig. 3.6 Superconducting qubit–oscillator circuit in deep-strong coupling regime. **a** Circuit diagram. A superconducting flux qubit (red and black) and a superconducting LC oscillator (blue and black) are inductively coupled to each other by sharing a tunable inductance (black). **b** Laser microscope image of the lumped-element LC oscillator inductively coupled to a coplanar transmission line. **c** Scanning electron microscope image of the qubit and the coupler junctions located at the red rectangle in image (**b**). The coupler, consisting of four parallel Josephson junctions, is tunable via the magnetic flux bias through its loops. (Reproduced with permission from Ref. [17].)

3.4 Rounding Up

It is no doubt that circuit quantum electrodynamics in the USC regime and beyond, opens up new frontiers that have never been realized before. Its applications in quantum computing are vast, and this thesis profits from this new interesting regime, with the hope of creating large-scale quantum computing architecture in mind. The rest of the thesis will focus mainly on that aspect.

Besides many interesting experimental phenomena resulting from the USC physics, it also touches upon profound insights in the ultimate nature of light–matter interaction. Let us recount this fascinating résumé. In physics, we use simplified models to understand/study what is really happening in Mother Nature, for instance, interactions between matter and electromagnetic fields in QED. Although such models have many assumptions and approximations, gauge invariance is generally preserved, which is the fundamental fact that physics is true for any inertia frame we are experiencing at. However, the choice of gauge has led to a historical dispute in the context of cavity QED, where the coupling of N qubits to a single radiation mode is described by the Dicke model [26]. This model predicts a superradiant phase transition [27, 28], when the collective light–matter coupling reaches the USC limit. It was shown later by a no-go theorem [29] that this transition does not show up when the "A^2-term" from the self-interaction energy of the field, in the underlying minimal coupling Hamiltonian is properly taken into account. But, by switching to the electric dipole gauge, this A^2-term can be eliminated [30, 31]. This shows clearly that models of light–matter interaction in different gauges give rise to the existence or nonexistence of a phase transition. Recently, the dispute has returned in superconducting qubits [32–34] and polaritons [35, 36].

Recently, it was shown that all these ambiguities can be fully resolved by a careful derivation and interpretation of the reduced effective cavity QED Hamiltonian [37]. All the recent articles [37–40] conclude that gauge ambiguities in QED arise due to the following factors: two-level approximation, and shape of the potential well. The validity of the two-level approximation for the dipoles explicitly matters on the gauge choice when the light–matter coupling becomes non-perturbative (USC). Hence, it is very important that one clearly knows the regimes at which the quantum Rabi model, especially, the two-level approximation in the model, is still valid. In following, we always stick to the regime where the quantum Rabi model works. In short, in an attempt to mimic cavity QED, we see that circuit QED is really something

> "More and Different."
>
> — P. W. Anderson.

References

1. Einstein A (1905) Über einen die erzeugung und verwandlung des lichtes betreffenden heuristischen gesichtspunkt. Annalen der physik 322(6):132
2. Kimble HJ (2008) The quantum internet. Nature 453:1023
3. Cohen-Tannoudji C, Dupont-Roc J, Grynberg G (1998) Atom-photon interactions: basic processes and applications. Wiley-VCH
4. Rabi II (1936) On the process of space quantization. Phys Rev 49:324
5. Braak D (2011) Integrability of the Rabi model. Phys Rev Lett 107:100401
6. Jaynes ET, Cummings FW (1963) Comparison of quantum and semiclassical radiation theories with application to the beam maser. Proc IEEE 51:89
7. Miller R, Northup TE, Birnbaum KM, Boca A, Boozer AD, Kimble HJ (2005) Trapped atoms in cavity QED: coupling quantized light and matter. J Phys B 38(9):S551
8. Kimble HJ (1998) Strong interactions of single atoms and photons in cavity QED. Phys Scr 1998(T76):127
9. Schoelkopf RJ, Girvin SM (2008) Wiring up quantum systems. Nature 451(7179):664
10. Thompson RJ, Rempe G, Kimble HJ (1992) Observation of normal-mode splitting for an atom in an optical cavity. Phys Rev Lett 68(8):1132
11. Wallraff A, Schuster DI, Blais A, Frunzio L, Huang R-S, Majer J, Kumar S, Girvin SM, Schoelkopf RJ (2004) Strong coupling of a single photon to a superconducting qubit using circuit quantum electrodynamics. Nature 431(7005):162
12. Blais A, Huang R-S, Wallraff A, Girvin SM, Schoelkopf RJ (2004) Cavity quantum electrodynamics for superconducting electrical circuits: an architecture for quantum computation. Phys Rev A 69:062320
13. Reed MD, DiCarlo L, Johnson BR, Sun L, Schuster DI, Frunzio L, Schoelkopf RJ (2010) High-fidelity readout in circuit quantum electrodynamics using the Jaynes-Cummings nonlinearity. Phys Rev Lett 105(17):173601
14. Forn-Díaz P, Lisenfeld J, Marcos D, García-Ripoll JJ, Solano E, Harmans CJPM, Mooij JE (2010) Observation of the Bloch-Siegert shift in a qubit-oscillator system in the ultrastrong coupling regime. Phys Rev Lett 105:237001
15. Niemczyk T, Deppe F, Huebl H, Menzel EP, Hocke F, Schwarz MJ, Garcia-Ripoll JJ, Zueco D, Hummer T, Solano E, Marx A, Gross R (2010) Circuit quantum electrodynamics in the ultrastrong-coupling regime. Nat Phys 6:772
16. Forn-Díaz P, Romero G, Harmans CJPM, Solano E, Mooij JE (2016) Broken selection rule in the quantum rabi model. Sci Rep 6:26720
17. Yoshihara F, Fuse T, Ashhab S, Kakuyanagi K, Saito S, Semba K (2017) Superconducting qubit-oscillator circuit beyond the ultrastrong-coupling regime. Nat Phys 13:44
18. Scalari G, Maissen C, Turčinková D, Hagenmüller D, De Liberato S, Ciuti C, Reichl C, Schuh D, Wegscheider W, Beck M, Faist J (2012) Ultrastrong coupling of the cyclotron transition of a 2D electron gas to a THz Metamaterial. Science 335(6074):1323
19. Gao W, Li X, Bamba M, Kono J (2018) Continuous transition between weak and ultrastrong coupling through exceptional points in carbon nanotube microcavity exciton-polaritons. Nat Photonics 12:362
20. Schwartz T, Hutchison JA, Genet C, Ebbesen TW (2011) Reversible switching of ultrastrong light-molecule coupling. Phys Rev Lett 106(19):196405
21. George J, Wang S, Chervy T, Canaguier-Durand A, Schaeffer G, Lehn J-M, Hutchison JA, Genet C, Ebbesen TW (2015) Ultra-strong coupling of molecular materials: spectroscopy and dynamics. Faraday Discuss 178:281
22. George J, Chervy T, Shalabney A, Devaux E, Hiura H, Genet C, Ebbesen TW (2016) Multiple rabi splittings under ultrastrong vibrational coupling. Phys Rev Lett 117(15):153601
23. Zhang X, Zou C-L, Jiang L, Tang HX (2014) Strongly coupled magnons and cavity microwave photons. Phys Rev Lett 113(15):156401
24. Forn-Díaz P, Lamata L, Rico E, Kono J, Solano E (2018) Ultrastrong coupling regimes of light-matter interaction. arXiv:1804.09275

25. Casanova J, Romero G, Lizuain I, García-Ripoll JJ, Solano E (2010) Deep strong coupling regime of the Jaynes-Cummings model. Phys Rev Lett 105:263603
26. Dicke RH (1954) Coherence in spontaneous radiation processes. Phys Rev 93(1):99
27. Hepp K, Lieb EH (1973) On the superradiant phase transition for molecules in a quantized radiation field: the Dicke maser model. Ann Phys 76(2):360
28. Wang YK, Hioe FT (1973) Phase transition in the Dicke model of superradiance. Phys Rev A 7(3):831
29. Rzażewski K, Wódkiewicz K, Żakowicz W (1975) Phase transitions, two-level atoms, and the A^2 term. Phys Rev Lett 35(7):432
30. Keeling J (2007) Coulomb interactions, gauge invariance, and phase transitions of the Dicke model. J Phys Condens Matter 19(29):295213
31. Vukics A, Grießer T, Domokos P (2014) Elimination of the A-square problem from cavity QED. Phys Rev Lett 112(7):073601
32. Nataf P, Ciuti C (2010) No-go theorem for superradiant quantum phase transitions in cavity QED and counter-example in circuit QED. Nat Commun 1:72
33. Viehmann O, von Delft J, Marquardt F (2011) Superradiant phase transitions and the standard description of circuit QED. Phys Rev Lett 107(11):113602
34. Jaako T, Xiang Z-L, Garcia-Ripoll JJ, Rabl P (2016) Ultrastrong-coupling phenomena beyond the Dicke model. Phys Rev A 94(3):033850
35. Chirolli L, Polini M, Giovannetti V, MacDonald AH (2012) Drude weight, cyclotron resonance, and the Dicke model of graphene cavity QED. Phys Rev Lett 109(26):267404
36. Hagenmüller D, Ciuti C (2012) Cavity QED of the graphene cyclotron transition. Phys Rev Lett 109(26):267403
37. De Bernardis D, Jaako T, Rabl P (2018a) Cavity quantum electrodynamics in the nonperturbative regime. Phys Rev A 97(4):043820
38. De Bernardis D, Pilar P, Jaako T, De Liberato S, Rabl P (2018b) Breakdown of gauge invariance in ultrastrong-coupling cavity QED. Phys Rev A 98:053819
39. Stokes A, Nazir A (2019) Gauge ambiguities imply Jaynes-Cummings physics remains valid in ultrastrong coupling QED. Nat Commun 10:499
40. Di Stefano O, Settineri A, Macrì V, Garziano L, Stassi R, Savasta S, Nori F (2018) Resolution of gauge ambiguities in ultrastrong-coupling cavity QED. arXiv:1809.08749

Chapter 4
Quantum Error-Correcting Codes in the USC Regime

There is no success without failure and losses.

—John C. Maxwell

In this chapter, we propose to construct relatively large quantum error-correcting codes (QECCs) with the superconducting circuits architecture at the ultrastrong coupling regime. We are able to create highly entangled quantum state commonly known as cluster state by generating entanglement between any pair of qubits within a fraction of a nanosecond. We coin this process "pairwise cluster state generation". To exemplify our proposal, the well-known five-qubit [1] and Steane codes [2] are numerically created. We construct them sequentially by performing ultrafast controlled-phase gates $\hat{U}_{CZ}$ between any two physical qubits. Ultrafast gate time and high fidelity response of the superconducting circuit might ensure very low errors incurred at logical qubit level (see Sect. 4.2). We believe our scheme could be used to mediate interactions between logical qubits and perform parity-protected quantum computations in a measurement-based manner [3]. In addition, our proposal may pave a way to construct various types of QECC applications [4, 5]. Lastly, we also provide optimal operating conditions with which more general graph codes could be realized with existing superconducting circuit technologies.

4.1 Pairwise Cluster State Generation

To achieve pairwise cluster state generation, let us consider a scenario depicted in Fig. 4.1a, where there are N identical flux qubits denoted by F's. They are uniformly distributed across a coplanar waveguide resonator (CWR). Two ends of the resonator are open, while the CWR supports one-dimensional current–charge waves with phase velocity $v = 1/\sqrt{lc}$ and wave impedance $Z_0 = \sqrt{l/c}$, where l and c are inductance and capacitance per unit length respectively (see Fig. 4.1b). Within

© Springer Nature Switzerland AG 2019
T. H. Kyaw, *Towards a Scalable Quantum Computing Platform in the Ultrastrong Coupling Regime*, Springer Theses,
https://doi.org/10.1007/978-3-030-19658-5_4

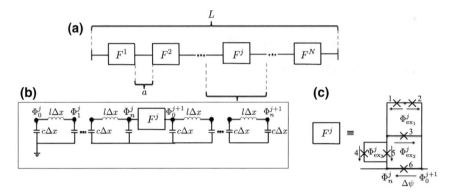

Fig. 4.1 a A coplanar waveguide resonator of length L interrupted by N identical and uniformly distributed flux qubits. **b** Lumped-element circuit model for a portion of the CWR, encompassing flux qubit F^j. **c** Schematic of the flux qubit. Numbers 1–6 with a cross sign label Josephson junctions, arrows refer to voltage drop between any two nodes, and $\Phi_{\mathrm{ex}1,2,3}$ are external magnetic fluxes passing through each loop

the lumped-element circuit formalism, dynamics or information about the CWR and the flux qubits are completely made known via the flux function, which we denote as $\Phi(x,t) = \int_{-\infty}^{t} V(x,t')dt'$, where $V(x,t)$ is an electrical potential of the CWR at position x with respect to the surrounding ground line. By following the circuit quantization recipe introduced in Chap. 2, the classical Lagrangian of the entire setup can be written as

$$\mathcal{L} = \sum_{j=1}^{N+1} \mathcal{L}_j^{\mathrm{CWR}} + \sum_{j=1}^{N} \mathcal{L}_j^{\mathrm{flux\ qubit}}, \tag{4.1}$$

with

$$\mathcal{L}_j^{\mathrm{CWR}} = \sum_{k=0}^{n} \frac{c\Delta x}{2}(\dot{\Phi}_k^j)^2 - \sum_{k=0}^{n-1} \frac{(\Phi_k^j - \Phi_{k+1}^j)^2}{2l\Delta x}, \quad\text{and} \tag{4.2}$$

$$\mathcal{L}_j^{\mathrm{flux\ qubit}} = \sum_{\ell=1}^{6} \frac{C_{J_\ell}}{2}\left(\frac{\Phi_0}{2\pi}\right)^2 (\dot{\varphi}_\ell^j)^2 + E_{J_\ell}\cos\left(\varphi_\ell^j\right). \tag{4.3}$$

Here, C_{J_ℓ} and E_{J_ℓ} are Josephson capacitance and Josephson energy of individual Josephson junction $(JJ)_\ell$ belonged to the jth flux qubit (see Fig. 4.1c). We also assume that $E_{J_1} = E_{J_2} = E_J$, $E_{J_3} = \alpha E_J$, $E_{J_4} = E_{J_5} = \beta E_J$ and $E_{J_6} = \gamma E_J$, where $\alpha, \beta, \gamma < 1$. Furthermore, φ_ℓ is the phase difference across JJ_ℓ, for example, $\varphi_6 = 2\pi(\Phi_0^{j+1} - \Phi_n^j)/\Phi_0$[1] and Δx is the lattice spacing of the lumped circuit element

---

[1] We remind ourselves that this is simply the Josephson phase–voltage relation.

description. With these system parameters and the flux quantization rule, we arrive at

$$
\mathcal{L}_j^{\text{flux qubit}} = \frac{C_J}{2}\left(\frac{\Phi_0}{2\pi}\right)^2\left[\left(\dot{\varphi}_1^j\right)^2 + \left(\dot{\varphi}_2^j\right)^2\right] + \frac{\alpha C_J}{2}\left(\frac{\Phi_0}{2\pi}\right)^2\left(\dot{\varphi}_2^j - \dot{\varphi}_1^j\right)^2
$$
$$
+ \beta C_J\left(\frac{\Phi_0}{2\pi}\right)^2\left(\dot{\varphi}_x^j + \dot{\varphi}_2^j - \dot{\varphi}_1^j\right)^2 + \frac{\gamma C_J}{2}\left(\frac{\Phi_0}{2\pi}\right)^2\left(\dot{\varphi}_x^j\right)^2
$$
$$
- U_q^j(\varphi_1^j, \varphi_2^j, \varphi_x^j), \tag{4.4}
$$

where the phase slip $\varphi_x^j = \Delta\psi^j = 2\pi(\Phi_n^j - \Phi_0^{j+1})/\Phi_0$ (see Fig. 4.1c) and

$$
\frac{U_q^j}{E_J} = - \left[\cos\varphi_1^j + \cos\varphi_2^j + \alpha\cos(\varphi_2^j - \varphi_1^j + 2\pi f_1^j)\right. \tag{4.5}
$$
$$
\left. + 2\beta\cos(\pi f_3^j)\cos(\varphi_2^j - \varphi_1^j + 2\pi(f_1^j - f_2^j + f_3^j/2) + \varphi_x^j)\right].
$$

where $f_k^j = \Phi_k/\Phi_0$ is a frustration parameter. When we consider the Kirchhoff's current law at the node Φ_n^j, the equation of motion is given by

$$
ca\ddot{\Phi}_n^j + (\gamma + 2\beta)C_J\left(\ddot{\Phi}_n^j - \ddot{\Phi}_0^{j+1}\right) + 2\beta C_J\left(\frac{\Phi_0}{2\pi}\right)\left(\ddot{\varphi}_2^j - \ddot{\varphi}_1^j\right)
$$
$$
= \frac{1}{l\Delta x}\left(\Phi_{n-1}^j - \Phi_n^j\right) - \gamma I_c\sin\varphi_x^j - \beta I_c\left[\sin(\varphi_x^j + \varphi_2^j - \varphi_1^j + 2\pi(f_1^j - f_2^j))\right.
$$
$$
\left. + \sin(\varphi_x^j + \varphi_2^j - \varphi_1^j + 2\pi(f_1^j - f_2^j + f_3^j))\right], \tag{4.6}
$$

where $I_c = 2\pi E_J/\Phi_0$. To arrive at a simple equation of motion, we make further assumption by stating that the Josephson inductance of JJ_6 in each flux qubit F^j is much smaller than the total inductance of each qubit loop, so that majority of the current flows through the resonator (a linear approximation of Josephson inductance [6]). In consequence, each flux qubit behaves like a small perturbation to the CWR. We then attain a simplified equation of motion

$$
ca\ddot{\Phi}_n^j + (\gamma + 2\beta)C_J\left(\ddot{\Phi}_n^j - \ddot{\Phi}_0^{j+1}\right) = \frac{1}{l\Delta x}\left(\Phi_{n-1}^j - \Phi_n^j\right) - \gamma I_c\sin\varphi_x^j, \tag{4.7}
$$

which represents the conservation of currents at the node Φ_n^j. This scenario has been thoroughly analyzed in Refs. [7, 8]. We thus decompose the Lagragian of JJ_6 into linear ($\mathcal{L}_j^{\text{JJ lin}}$) and nonlinear ($\mathcal{L}_j^{\text{JJ nonlin}}$) components as

$$
\mathcal{L}_j^{\text{JJ lin}} = (\gamma + 2\beta)\frac{C_J}{2}\left(\dot{\Phi}_n^j - \dot{\Phi}_0^{j+1}\right)^2 - \frac{1}{2L_J}\left(\Phi_n^j - \Phi_0^{j+1}\right)^2, \text{ and} \tag{4.8}
$$

$$\mathcal{L}_j^{\text{JJ nonlin}} = \gamma E_J \cos\left(\frac{2\pi(\Phi_n^j - \Phi_0^{j+1})}{\Phi_0}\right) + \frac{1}{2L_J}\left(\Phi_n^j - \Phi_0^{j+1}\right)^2. \qquad (4.9)$$

We have introduced the second terms in Eqs. (4.8) and (4.9), to remove nonlinear component on purpose. In the continuum limit $\Delta x \to 0$, we arrive at

$$\mathcal{L}_j^{\text{CWR}} = \int_{(j-1)a}^{ja}\left\{\frac{ca}{2}[\partial_t\Phi(x,t)]^2 - \frac{1}{2la}[\partial_x\Phi(x,t)]^2\right\}dx, \qquad (4.10)$$

where $a = L/(N + 1)$ is the lattice spacing between junctions JJ_6 (see Fig. 4.1a). Furthermore, we have

$$\mathcal{L}_j^{\text{JJ lin}} = (\gamma + 2\beta)\frac{C_J}{2}\delta\dot{\Phi}_j^2 - \frac{1}{2L_J}\delta\Phi_j^2, \text{ and} \qquad (4.11)$$

$$\mathcal{L}_j^{\text{JJ nonlin}} = \gamma E_J \cos\delta\varphi_j + \frac{1}{2L_J}\delta\Phi_j^2, \qquad (4.12)$$

where $\delta\Phi_j = \Phi|_{x\to ja^-} - \Phi|_{x\to ja^+} = \frac{\Phi_0}{2\pi}\varphi_x^j$ is the flux drop across JJ_6 of the jth flux qubit, in the limits of the flux approaching the JJ_6 from its left side ($\Phi|_{x\to ja^-}$) and from the right ($\Phi|_{x\to ja^+}$). By considering the following three factors: (1) the boundary conditions of vanishing currents at the two end points of the CWR, which we have assumed initially: $\partial_x\Phi|_{x=0} = \partial_x\Phi|_{x=L} = 0$, (2) the conservation of currents at each JJ_6: $\partial_x\Phi|_{x\to ja^-} = \partial_x\Phi|_{x\to ja^+}$, and (3) the JJ_6 current–flux relationship: $-\partial_x\Phi|_{x=ja}/l = (\gamma + 2\beta)C_J\delta\ddot{\Phi}_j + \delta\Phi_j/L_J$, we arrive at a well-defined eigenvalue problem [8]. One can transform the linear part of the JJ_6's doped CWR into independent harmonic oscillators [7, 8] with solutions of the eigenmode functions. After performing a Legendre transformation, we arrive at the full Hamiltonian

$$\mathcal{H}_{\text{CWR}} = \sum_i^\infty \frac{1}{2m_i}\pi_i^2 + \frac{1}{2}m_i^2\omega_i^2\tau_i^2 + \mathcal{H}_{NL}, \qquad (4.13)$$

where $m_i = c\int_0^L r_i^2 dx + (\gamma + 2\beta)C_J\sum_{j=1}^N(r_i|_{x\to ja^-} - r_i|_{x\to ja^+})^2$ is the effective mass of the ith eigenmode [7], $\pi_i = m_i\dot{\tau}_i$ is the canonical conjugate momentum of τ_i and $\mathcal{H}_{NL} = \frac{-1}{L_J}\left(\frac{\Phi_0}{2\pi}\right)^2\sum_{j=1}^N\left[\gamma\cos(\delta\varphi_j) + \left(\frac{2\pi}{\Phi_0}\right)^2\frac{\delta\Phi_j^2}{2}\right]$ is the nonlinear Hamiltonian, coming from the Lagrangian (nonlinear), Eq. (4.12). Here, we assume an ansatz for the flux function $\phi(x,t) = \sum_i \tau_i(t)r_i(x)$. By imposing the canonical commutation relation- $[\hat{\pi}_n, \hat{\tau}_m] = -i\delta_{nm}\hat{I}$, we quantize the theory with annihilation (creation) operators $\hat{a}_i = \sqrt{m_i\omega_i/(2)}(\hat{\tau}_i + i\hat{\pi}_i/(m_i\omega_i))$ $(\hat{a}_i^\dagger)$. From here onwards, we use the unit where $\hbar = 1$. Henceforth, we can finally introduce operator notion in our classical Hamiltonians as such $\hat{\mathcal{H}}_{\text{CWR}} = \sum_i \omega_i\hat{a}_i^\dagger\hat{a}_i + \hat{\mathcal{H}}_{NL}$, where $\omega_i = (\pi v/L)(N + 1)m$

with $m \in \mathbb{N}$, and $\nu = 1/\sqrt{lc}$. We further impose that each JJ$_6$ operates in a linear approximation of Josephson inductance [6] such that $\hat{\mathcal{H}}_{NL} \approx \hat{0}$.

In a single-band approximation or a plasma frequency $\omega_p = \bar{\omega}$ with $\bar{\omega} = \pi v(N + 1)/L$, when a set of $\mathcal{M}$ eigenmodes become degenerate [8], we have $\hat{\mathcal{H}}_{CWR} = \sum_i \omega_i \hat{a}_i^\dagger \hat{a}_i$. In order to obtain the jth flux qubit energy and the qubit–resonator coupling, we expand the qubit potential term, Eq. (4.5), up to the first order in $\hat{\varphi}_x^j$ [9], leading to the approximated potential energy

$$\hat{U}_q^j/E_J \approx - [\cos \hat{\varphi}_1^j + \cos \hat{\varphi}_2^j + \alpha \cos(\hat{\varphi}_2^j - \hat{\varphi}_1^j + 2\pi f_1^j) \qquad (4.14)$$
$$+ 2\beta \cos(\pi f_3^j) \cos(\hat{\varphi}_2^j - \hat{\varphi}_1^j + 2\pi(f_1^j - f_2^j + f_3^j/2))]$$
$$+ 2\beta \cos(\pi f_3^j) \sin(\hat{\varphi}_2^j - \hat{\varphi}_1^j + 2\pi(f_1^j - f_2^j + f_3^j/2))\hat{\varphi}_x^j.$$

The first and second components in $\hat{U}_q^j/E_J$ determine the flux qubit potential, while the third one is the qubit–resonator coupling. The flux qubit Hamiltonian is then obtained by summing up the kinetic energy terms appearing in Eq. (4.4) and the first two terms of Eq. (4.14), followed by a Legendre transformation. At the symmetry point, i.e., $\Phi_1 \approx \Phi_0/2$, the flux qubit potential (4.14) can be effectively truncated to a two-level system with frequency $\omega_q = \sqrt{\Delta^2 + \varepsilon^2}$. Here, Δ is the qubit energy gap and $\varepsilon = 2I_p(f_{ex_1} - 1/2)\Phi_0$ with I_p being the persistent current. The qubit–resonator coupling is also obtained by projecting the operator $\sin(\hat{\varphi}_2^j - \hat{\varphi}_1^j + 2\pi(f_1^j - f_2^j + f_3^j/2))$ into the qubit basis, that is

$$\sin(\hat{\varphi}_2^j - \hat{\varphi}_1^j + 2\pi(f_1^j - f_2^j + f_3^j/2)) = \sum_{\nu=0,x,y,z} c_\nu \hat{\sigma}_\nu, \qquad (4.15)$$

where $\hat{\sigma}_0 = \hat{I}$ being the identity operator, and c_ν are c-numbers obtained numerically. Hence, the Hamiltonian of the overall setup $\hat{\mathcal{H}} = \hat{\mathcal{H}}_{CWR} + \hat{\mathcal{H}}_{\text{flux qubits}} + \hat{\mathcal{H}}_{\text{interaction}}$ (c.f. Eq. 4.1) becomes

$$\hat{\mathcal{H}} = \frac{1}{2}\sum_{j=1}^{N} \omega_q^j \hat{\sigma}_z^j + \sum_{\ell \in \mathcal{M}} \omega_\ell \hat{a}_\ell^\dagger \hat{a}_\ell + \sum_{j=1}^{N}\sum_{\ell \in \mathcal{M}} g_j(c_x^j \hat{\sigma}_x^j + c_z^j \hat{\sigma}_z^j)(\hat{a}_\ell + \hat{a}_\ell^\dagger), \qquad (4.16)$$

which is the starting point of our proposal. An artistic impression of the setup is shown in Fig. 4.2a. Here, ω_q^j is the jth qubit frequency, $\hat{\sigma}_{z,x}^j$ are the Pauli matrices, ω_l is the frequency of the ℓth resonator mode belonging to the manifold $\mathcal{M}$, $a_\ell^\dagger(a_\ell)$ is the creation (annihilation) operator of ℓth resonator mode, and the coefficients c_x^j and c_z^j are functions of the system parameters α, β, f_1, and f_2 [9], satisfying the condition $|c_x^j|^2 + |c_z^j|^2 = 1$ for $\forall j$. The coupling strength $g_j = 2\beta E_J \frac{\Phi_0}{2\pi} \cos(\pi f_3)\Delta\psi^j$ are the effective coupling strengths between the jth flux qubit and CWR at the degeneracy point with $\Delta\psi^j \propto \sqrt{2/(N + 1)} \sin(p_j)$ with $p_j = \pi j/(N+1)$. In particular, it depends on the external magnetic flux Φ_{ex_3} threading each flux qubit (see Fig. 4.2), and N is the total number of qubits present in the resonator. We note that different coupling

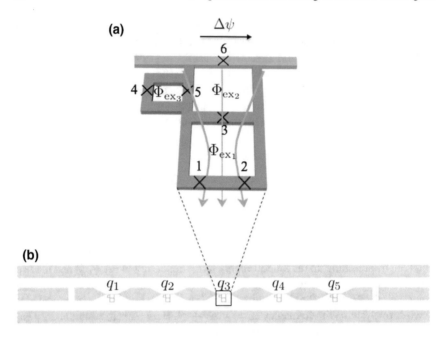

Fig. 4.2 **a** Schematic of a flux qubit, denoted by the Josephson junctions 1, 2, and 3. By varying the frustration parameter f_3, attained by an applied magnetic flux passing through the loop composed of the JJ$_4$ and JJ$_5$, the coupling between the qubit and the resonator can be tuned at will. This is a crucial aspect of our superconducting qubits design in order to realize cluster states in an ultrafast timescale. **b** An array of five USC qubits embedded in a resonator to obtain the five-qubit quantum error-correcting code

strengths appear due to the spatial dependence of the distribution of flux qubits along the resonator. Moreover, it has been shown that different frequencies belonging to a specific manifold $\mathcal{M}$ become degenerate ($\omega_\ell = \omega$) [8] for a specific value of the plasma frequency $\omega_p = 1/\sqrt{C_J L_J}$ associated with the coupling junctions JJ$_6$.

It is noteworthy that coefficients c_x^j and c_z^j can be manipulated by means of the external flux $\Phi_{ex_1}^j$ as shown in Fig. 4.3a, b for a given junction size $\alpha \doteq E_{J_3}/E_J$. Here, it might be possible to tune the transversal coupling where $c_x \to 1$ and $c_z \to 0$, or the longitudinal coupling where $c_x \to 0$ and $c_z \to 1$. The latter becomes an essential condition for generating pairwise cluster states. The numerical simulation of coefficients c_x and c_z in Fig. 4.3a, b has been performed by diagonalizing the flux qubit potential, Eq. (4.14), and truncating it to the two lowest energy levels [9].

A cluster state between any ith and jth qubits can be readily generated in four steps. First, the system is cooled down to its USC ground state. Second, both qubits are adiabatically tuned with external magnetic fluxes. The latter vary linearly in time[2]

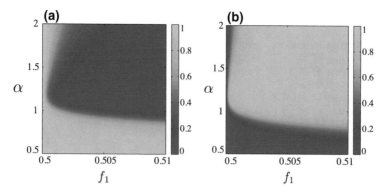

Fig. 4.3 Coupling coefficients **a** c_x and **b** c_z in Eq. (4.16) as a function α, that is the size of junction JJ$_3$, and the frustration parameter $f_1 = f_{ex_1}$. In this simulation, we have considered $E_J/h = 221$ GHz and the capacitive energy of junctions as $E_C = E_J/32$

$\Phi^k_{ex_3} = \bar{\phi}_0 + (\Delta\phi)t/T$ where $k \in \{i, j\}$, with $\bar{\phi}_0$ being an offset flux, $\Delta\phi$ being a small flux amplitude, and T being the total evolution time. In this case, the coupling strength of each qubit reaches the strong coupling regime described by the Jaynes–Cummings model [10] such that the system is prepared in the state $|\psi_G\rangle = |g\rangle^{\otimes N} \otimes |0\rangle^{\otimes N}$, where $|g\rangle$ and $|0\rangle$ stand for the ground state of the qubit and the vacuum state for each mode in $\mathcal{M}$, respectively. Third, each qubit is then addressed with a classical microwave signal, sent through the cavity, to be prepared in the state $|+\rangle = (|g\rangle + |e\rangle)/\sqrt{2}$ which is an eigenstate of σ_x, while all the remaining $N - 2$ qubits are far off-resonant with respect to the driving frequency. At this stage, all qubits should dispersively interact with the modes within the manifold $\mathcal{M}$ such that there is no exchange of excitations. This task might be carried out at a degenerate regime of the bosonic manifold, $\omega_\ell = \omega$ [8]. At last, the external magnetic fluxes $\Phi^k_{ex_3}$ are swiftly tuned to reach the USC coupling strength within a subnanosecond timescale. During the four-step process, the magnetic fluxes $\Phi^k_{ex_1}$ should be tuned to reach the longitudinal qubit–resonator coupling.

To illustrate our protocol, we consider two qubits embedded in a resonator with two modes and simulate the aforementioned adiabatic process. Figure 4.4 shows the fidelity plot of the JC ground state $|\psi_{JC}\rangle = |gg\rangle \otimes |00\rangle$ and the instantaneous state $|\psi(t)\rangle$ during an adiabatic switch off process, given the initial state $|\psi_G\rangle$, that is, the ground state of the quantum Rabi model. The initialization process via adiabatic switch off takes $T = 250/\omega = 50$ ns provided the resonator frequency is $\omega = 5$ GHz. Obtaining unit fidelity at the end of the adiabatic evolution ascertains that the rotating wave approximation is consistent in this context. The extension to a large number of qubits and bosonic modes is straightforward.

After interacting with the collective resonator modes, the system evolution operator takes the form [11] (see Appendix A for the detailed derivation).

---

[2]We choose the linear magnetic sweep for simplicity. Other tuning scheme is equally valid as long as the adiabatic theorem is satisfied.

Fig. 4.4 Fidelity between the desired Jaynes–Cummings ground state $|\psi_{JC}\rangle = |gg\rangle \otimes |00\rangle$ and the instantaneous state $|\psi(t)\rangle$ during an adiabatic switch off process, provided that an initial state of the evolution is $|\psi_G\rangle$, ground state of the quantum Rabi model. In another words, $\mathcal{F} = |\langle \psi_{JC}|\psi(t)\rangle|^2$ is plotted against g/ω

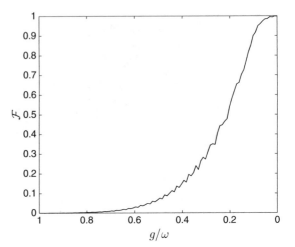

$$\hat{U}(t) = \hat{U}_0(t) e^{i\hat{\xi}^2 M (\omega t - \sin(\omega t))} \prod_\ell e^{-i\omega t \hat{a}_\ell^\dagger \hat{a}_\ell} \hat{\mathcal{D}}_\ell[\kappa(t)], \qquad (4.17)$$

where $\hat{\mathcal{D}}_\ell[\kappa(t)] = \exp[\kappa(t)\hat{a}_\ell^\dagger - \kappa^*(t)\hat{a}_\ell]$ is the displacement operator associated with the ℓth bosonic mode within the manifold $\mathcal{M}$. In addition, $\hat{\xi} = \sum_{j=1}^N \kappa_j \hat{\sigma}_z^j$ with $\kappa_j = g_j/\omega$, M stands for the number of degenerate bosonic modes $\hat{a}_\ell$, and the unitary $\hat{U}_0(t) = \exp(-it \sum_{j=1}^N \frac{\omega_q^j}{2}\hat{\sigma}_z^j)$. After the evolution time $T = 2\pi n/\omega$, we have performed the desired controlled-phase gate operation between the qubits

$$\hat{U}_{CZ}^{ij} = \hat{\mathcal{U}} \times \exp\left[-\frac{i\pi}{4}(\hat{\sigma}_z^i + \hat{\sigma}_z^j)\right] \qquad (4.18)$$

$$\times \exp\left[4\pi i M \left((\kappa_i^2 + \kappa_j^2)\frac{\hat{I}}{2} + \kappa_i \kappa_j \hat{\sigma}_z^i \hat{\sigma}_z^j\right)\right],$$

where $\hat{\mathcal{U}} = \exp\left[\frac{-i\pi}{4}\left[\left(\frac{4\omega_q^i - \omega}{\omega}\right)\hat{\sigma}_z^i + \left(\frac{4\omega_q^j - \omega}{\omega}\right)\hat{\sigma}_z^j\right]\right]$. The resultant state incurs an extra global phase due to the presence of $\hat{\mathcal{U}}$, which is unavoidable since it is formidable by construction to tune a desired qubit frequency, via the external flux Φ_{ex_1}, without affecting the longitudinal and transversal coupling strengths (see Fig. 4.3a and b). To achieve maximum gate fidelity, we require both $\kappa_i^2 + \kappa_j^2 = \frac{1}{8nM}$ and $\kappa_i \kappa_j = \frac{1}{16nM}$. That means the two coupling strengths need to satisfy $\kappa_i + \kappa_j = \frac{1}{2\sqrt{nM}}$. The operational gate time is estimated to be $T = 2\pi/\omega \sim 0.2$ ns if the collective mode frequency is $\omega = 2\pi \times 5$ GHz, which implies a ratio $g_j/\omega = 1/(4\sqrt{2}) \approx 0.17$ operating at the USC regime. As soon as the two qubits are entangled, they are immediately detuned from the resonant frequency so that we may repeat the same procedure for other

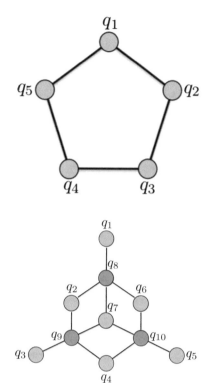

Fig. 4.5 The five-qubit cluster state in a ring geometry. Each qubit represented by q_i is prepared in $|+\rangle$ state, and each black bond represents a controlled-phase gate operation

Fig. 4.6 A ten-qubit cluster state giving rise to the Steane code, after appropriate projective measurements on the physical qubits q_8, q_9 and q_{10} (orange circles). Each black bond represents the pairwise cluster state generation mechanism U_{CZ}^{ij} and each circle is initially prepared in the $|+\rangle = (|0\rangle + |1\rangle)/\sqrt{2}$ state

qubit pairs to arrive at a specific quantum error-correcting code, be it the five-qubit code (see Fig. 4.5) or the Steane code (see Fig. 4.6).

4.1.1 Five-Qubit Code

To demonstrate our scheme, we create the five-qubit code which is the smallest QECC that protects against an arbitrary quantum error on a single qubit [1]. Let us explain a little bit more what we mean by "an arbitrary quantum error on a single qubit". Suppose that our encode qubit state is $|\psi\rangle = \alpha|0_L\rangle + \beta|1_L\rangle$, where $|0/1_L\rangle$ are the logical qubits, composing of five physical qubits in the present case. With some noise acting on the state, we then have $\mathcal{E}(|\psi\rangle\langle\psi|) = \sum_i \hat{E}_i|\psi\rangle\langle\psi|\hat{E}_i^\dagger$, where $\mathcal{E}$ is a trace-preserving quantum operation with $\sum_k \hat{E}_k^\dagger \hat{E}_k = \hat{I}$. The arbitrary quantum error means $\hat{E}_i = e_{i0}\hat{I}^j + e_{i0}\hat{\sigma}_x^j + e_{i2}\hat{\sigma}_z^j + e_{i3}\hat{\sigma}_x^j\hat{\sigma}_z^j$, acting on the jth qubit among the five qubits. Furthermore, in the stabilizer formalism, the five-qubit code is the common eigenspace of the following stabilizer operators shown in Table 4.1. It is easy to show that the five-qubit logical qubits are $|0_L\rangle = \sum_{i=1}^4 \hat{M}_i|00000\rangle$ and $|1_L\rangle = \bar{X}|0_L\rangle$, where $\hat{M}_i$ and $\bar{X}$ are listed in Table 4.1. In the following, we will try to show that the five-

Table 4.1 Stabilizer generators for the five-qubit code, its logical $\bar{Z}$ and logical $\bar{X}$ operators

Name	1	2	3	4	5
$\hat{M}_1$	$\hat{\sigma}_x$	$\hat{\sigma}_z$	$\hat{\sigma}_z$	$\hat{\sigma}_x$	$\hat{I}$
$\hat{M}_2$	$\hat{I}$	$\hat{\sigma}_x$	$\hat{\sigma}_z$	$\hat{\sigma}_z$	$\hat{\sigma}_x$
$\hat{M}_3$	$\hat{\sigma}_x$	$\hat{I}$	$\hat{\sigma}_x$	$\hat{\sigma}_z$	$\hat{\sigma}_z$
$\hat{M}_4$	$\hat{\sigma}_z$	$\hat{\sigma}_x$	$\hat{I}$	$\hat{\sigma}_x$	$\hat{\sigma}_z$
$\bar{X}$	$\hat{\sigma}_x$	$\hat{\sigma}_x$	$\hat{\sigma}_x$	$\hat{\sigma}_x$	$\hat{\sigma}_x$
$\bar{Z}$	$\hat{\sigma}_z$	$\hat{\sigma}_z$	$\hat{\sigma}_z$	$\hat{\sigma}_z$	$\hat{\sigma}_z$

qubit code stabilizer generators are related to stabilizers of the five-qubit cluster state connected in a ring geometry as shown in Fig. 4.5.

A cluster state is defined as a common eigenstate of stabilizer operators $\hat{K}_i = \hat{\sigma}_x^i \bigotimes_{j\in nb(i)} \hat{\sigma}_z^j$ with +1 eigenvalue. Here, $nb(i)$ means connected neighbors of the ith qubit. All the stabilizers associated to the cluster state shown in Fig. 4.5 are listed in Table 4.2. In addition, $|\psi\rangle = \hat{K}_i|\psi\rangle = \hat{K}_i\hat{K}_j|\psi\rangle$, where $|\psi\rangle$ is a cluster state. Hence, one can define a new set of operators $\hat{S}'_i = \hat{K}_i\hat{K}_{i+1 \bmod 5}$ and logical operators $\hat{\bar{X}} = \hat{K}_5$ and $\hat{\bar{Z}} = \hat{Z}_1\hat{Z}_2\hat{Z}_3\hat{Z}_4\hat{Z}_5$, from which it follows that the five-qubit cluster state is equivalent to the five-qubit QECC via local unitary transformation $\hat{U} = \bigotimes_i \hat{S}_i\hat{H}_i$, where $\hat{S}_i$ ($\hat{H}_i$) is the phase (Hadamard) gate. The relationship is obvious by comparing the two Tables 4.1 and 4.2. We remark that this kind of cluster state is also used for concatenated quantum codes [12]. Since we have shown their equivalence, we now turn to create a five-qubit cluster state in the ring geometry shown in Fig. 4.5 via the pairwise cluster state generation procedure $\left(\hat{U}_{CZ}^{ij}\right)$ proposed in Sect. 4.1. The resultant cluster state is

$$|\Psi_5\rangle = \hat{U}_{CZ}^{15}\hat{U}_{CZ}^{54}\hat{U}_{CZ}^{43}\hat{U}_{CZ}^{32}\hat{U}_{CZ}^{21}|+\rangle^{\otimes 5}, \tag{4.19}$$

Table 4.2 Stabilizer operators associated with the five-qubit cluster state in a ring as shown in Fig. 4.5, its logical $\bar{Z}$ and logical $\bar{X}$ operators

Name	1	2	3	4	5
$\hat{S}'_1$	$\hat{\sigma}_y$	$\hat{\sigma}_y$	$\hat{\sigma}_z$	$\hat{I}$	$\hat{\sigma}_z$
$\hat{S}'_2$	$\hat{\sigma}_z$	$\hat{\sigma}_y$	$\hat{\sigma}_y$	$\hat{\sigma}_z$	$\hat{I}$
$\hat{S}'_3$	$\hat{I}$	$\hat{\sigma}_z$	$\hat{\sigma}_y$	$\hat{\sigma}_y$	$\hat{\sigma}_z$
$\hat{S}'_4$	$\hat{\sigma}_z$	$\hat{I}$	$\hat{\sigma}_z$	$\hat{\sigma}_y$	$\hat{\sigma}_y$
$\hat{K}_5 \equiv \bar{X}$	$\hat{\sigma}_z$	$\hat{I}$	$\hat{I}$	$\hat{\sigma}_z$	$\hat{\sigma}_x$
$\bar{Z}$	$\hat{\sigma}_z$	$\hat{\sigma}_z$	$\hat{\sigma}_z$	$\hat{\sigma}_z$	$\hat{\sigma}_z$

after an evolution time $\tau_5 = 5 \times 2\pi/\omega$. Finally, we achieve the five-qubit code after local operations have acted on the individual qubits. Here, the numeral superscripts represent the location of physical qubits within the superconducting circuit shown in Fig. 4.2b.

4.1.2 Steane Code

Another QECC we want to generate within our scheme is the Steane code [2], whose stabilizer generators are listed in Table 4.3.

The code is not equivalent to a cluster state. However, it can be obtained from the graph shown in Fig. 4.6, where each circle represents a qubit initialized in $|+\rangle$ state and each bond represents a controlled-phase gate. It is easy to show that after measuring the orange qubits in the X basis, the remaining seven qubits is equivalent to the code state [1], which is nothing but $|0_L\rangle = \sum_{i=1}^{7} \hat{M}_i |0\rangle^{\otimes 7}$, and $|1_L\rangle = \bar{X}|0_L\rangle$. This state will depend on the outcome of the measurement. Here, $\hat{M}_i$ and $\bar{X}$ are listed in Table 4.3. This translates to twelve $\hat{U}_{CZ}^{ij}$ gates followed by three parallel measurements within an evolution time of $\tau_7 = 12 \times 2\pi/\omega$, we would achieve the Steane code-

$$|\Psi_7\rangle = \langle +|_{10}\langle +|_9\langle +|_8 \prod_{k \in E} \hat{U}_{CZ}^k |+\rangle^{\otimes 10} \tag{4.20}$$

with E representing the set of all the black-colored bonds in Fig. 4.6.

Table 4.3 Stabilizer generators for the Steane code, its logical $\bar{Z}$ and logical $\bar{X}$ operators

Name	1	2	3	4	5	6	7
$\hat{M}_1$	$\hat{\sigma}_x$	$\hat{\sigma}_x$	$\hat{I}$	$\hat{I}$	$\hat{I}$	$\hat{\sigma}_x$	$\hat{\sigma}_x$
$\hat{M}_2$	$\hat{\sigma}_z$	$\hat{\sigma}_z$	$\hat{I}$	$\hat{I}$	$\hat{I}$	$\hat{\sigma}_z$	$\hat{\sigma}_z$
$\hat{M}_3$	$\hat{I}$	$\hat{\sigma}_x$	$\hat{\sigma}_x$	$\hat{\sigma}_x$	$\hat{I}$	$\hat{I}$	$\hat{\sigma}_x$
$\hat{M}_4$	$\hat{I}$	$\hat{\sigma}_z$	$\hat{\sigma}_z$	$\hat{\sigma}_z$	$\hat{I}$	$\hat{I}$	$\hat{\sigma}_z$
$\hat{M}_5$	$\hat{I}$	$\hat{I}$	$\hat{I}$	$\hat{\sigma}_x$	$\hat{\sigma}_x$	$\hat{\sigma}_x$	$\hat{\sigma}_x$
$\hat{M}_6$	$\hat{I}$	$\hat{I}$	$\hat{I}$	$\hat{\sigma}_z$	$\hat{\sigma}_z$	$\hat{\sigma}_z$	$\hat{\sigma}_z$
$\bar{X}$	$\hat{\sigma}_x$	$\hat{\sigma}_x$	$\hat{\sigma}_x$	$\hat{\sigma}_x$	$\hat{\sigma}_x$	$\hat{\sigma}_x$	$\hat{\sigma}_x$
$\bar{Z}$	$\hat{\sigma}_z$	$\hat{\sigma}_z$	$\hat{\sigma}_z$	$\hat{\sigma}_z$	$\hat{\sigma}_z$	$\hat{\sigma}_z$	$\hat{\sigma}_z$

4.2 Errors and Decoherence Model

4.2.1 Nonzero Transversal Light–Matter Coupling

We would like to prepare our QECC within a microwave cavity in a fast timescale. Here, we assume that one can embed a large number of qubits inside the resonator line and the light–matter coupling between individual two-level system and the resonator can be tuned easily via an external magnetic field without any possible crosstalk (which is not true in reality and we will devote the entire Chap. 6 to address the issue). Even in an ideal case, our proposed pairwise cluster state generation mechanism assumes coupling coefficients $c_x^j = 0$ in the effective Hamiltonian, Eq. (4.16),

$$\hat{\mathcal{H}} = \frac{1}{2} \sum_{j=1}^{N} \omega_q^j \hat{\sigma}_z^j + \sum_{\ell \in \mathcal{M}} \omega_\ell \hat{a}_\ell^\dagger \hat{a}_\ell + \sum_{j=1}^{N} \sum_{\ell \in \mathcal{M}} g_j (c_x^j \hat{\sigma}_x^j + c_z^j \hat{\sigma}_z^j)(\hat{a}_\ell + \hat{a}_\ell^\dagger),$$

i.e., we desire that only longitudinal couplings are present in the setup. However, there might be some residual nonzero transversal couplings in an experimental implementation. Whenever $c_x^j \neq 0$, the ultrafast gate $\hat{U}_{CZ}^{ij}$ performance is unavoidably affected. In order to benchmark some acceptable threshold at which the performance of our quantum gate is still acceptable with the presence of nonzero transversal couplings, we perform numerical simulations for the dynamics governed by Eq. (4.16) for the simplest scenario of two qubits and two bosonic modes belonging to the manifold $\mathcal{M}$. In Fig. 4.7, we show the optimal operating conditions to obtain maximum gate fidelity. In particular, we plot the fidelity $\mathcal{F} = |\langle \psi_F | \psi \rangle|^2$ between the expected final two-qubit state $|\psi_F\rangle = \frac{1}{\sqrt{2}}(|e, +\rangle - |g, -\rangle)$, with $|\pm\rangle = (|g\rangle \pm |e\rangle)/\sqrt{2}$, and the state $|\psi\rangle$ after the pairwise gate operation has been performed with an initial state $|\psi_0\rangle = |+, +\rangle\langle+, +| \otimes \hat{\rho}_{cav}$ along various c_x values. We then obtain the fidelity $\mathcal{F}$ after tracing out the resonator degrees of freedom. Here, $\hat{\rho}_{cav}$ is the cavity field being a thermal state at 15 mK (red solid line), a vacuum state (black solid line), a coherent state $|\alpha\rangle$ with the amplitude $\alpha = 0.25$ (blue solid line), $\alpha = 0.5$ (dashed line), and $\alpha = 1$ (dotted line), respectively. Even though the presence of vacuum, thermal or coherent state inside the resonator at near $c_x \ll 1$ does not affect much of the gate performance, we note that a coherent state field in the resonator has clear advantage over the true vacuum field. In particular, we observe improvement in the gate fidelity when the resonator field becomes closer to the classical field, i.e., for a coherent state amplitude $\alpha \to 1$.

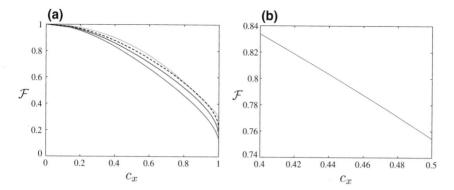

Fig. 4.7 **a** Fidelity of achieving the desired controlled-phase gate $\hat{U}_{CZ}$, in the presence of nonzero transversal coupling strength c_x, while the red solid, the black solid, the blue solid, the dashed, and the dotted lines represent the case when the resonator field is a thermal state at 15 mK, a vacuum state, a coherent state $|\alpha\rangle$ with the amplitude $\alpha = 0.25$, $\alpha = 0.5$ and $\alpha = 1$, respectively. **b** The enlarged figure of **a** shows the red and black solid lines overlap each other, indicating that the vacuum and the thermal states behave the same when considering two bosonic modes

4.2.2 Decoherence Noise Modeling via Monte Carlo Simulation

In addition to imperfection of the resonator initial state and the qubit–resonator coupling strengths, we expect our system to be exposed to thermal noise from the control lines and crosstalks between physical qubits. We assume that those noises occurring in physical qubits are uncorrelated, which is a conservative assumption. A rough estimate of the error per quantum gate is

$$p_{\text{error}}^{\text{gate}}(t, T) = \Gamma_{\text{gate}}(t, T) \times t, \tag{4.21}$$

where we see that p depends on the decoherence lifetime Γ of a physical qubit, which one has to model and calculate from the microscopic derivation of the USC system and bath (see Chap. 5), as well as the each gate operation time t. T refers to the total time to perform all the gates needed to perform a task. Furthermore, there are four transitions the system undergoes in terms of quantum computing paradigm. They are:

1. Quantum state preparation,
2. System idle time,
3. Gate operations (single- and two-qubit), and
4. State measurements.

All the four processes are subject to noise. In the following, we model each of the four processes (except the system idle time) induced by the possible noises $\sigma_x, \sigma_y, \sigma_z$

during the QECC generation, in terms of a simple Monte Carlo algorithm outlined
below.

```
function ERROR(Gate1,perror)
    prob = rand
    if prob < perror then
        prob2 = rand
        if prob2 < 1/3 then return Gate2 = X × Gate1
        else if 1/3 ≤ prob2 < 2/3 then return Gate2 = Y × Gate1
        else   return Gate2 = Z × Gate1
        end if
    else   return Gate2 = Gate1
    end if
end function
```

In short, for the state preparation, one can randomly incur noise by calling the error
function (the Monte Carlo algorithm above) such that $\text{ERROR}(\hat{I}, p_{\text{error}})|+\rangle$, which
is performed as many as the number of qubits present in the code. The two-qubit
controlled-phase gate (Cz) can be represented in terms of Pauli matrices as

$$Cz_{ij} = \frac{1}{2}\left((\hat{I} + \hat{\sigma}_z)_i \otimes \hat{I}_j + (\hat{I} - \hat{\sigma}_z)_i \otimes \hat{\sigma}_{zj}\right). \tag{4.22}$$

For the erroneous Cz, similar to Eq. (4.22), one has

$$Cze_{ij} = \frac{1}{2}\bigg((\text{ERROR}(\hat{I}, p_{\text{error}}) + \text{ERROR}(\hat{\sigma}_z, p_{\text{error}}))_i \otimes (\text{ERROR}(\hat{I}, p_{\text{error}}))_j +$$
$$(\text{ERROR}(\hat{I}, p_{\text{error}}) - \text{ERROR}(\hat{\sigma}_z, p_{\text{error}}))_i \otimes (\text{ERROR}(\hat{\sigma}_z, p_{\text{error}}))_j\bigg). \tag{4.23}$$

The same applies to quantum state measurements. This is needed especially in the
Steane code generation Fig. 4.6, but with a constant measurement error $p_m = 0.01^3$
[13]. At the end, the collective state of the logical qubit associated with the graph code,
be it the five-qubit code or the Steane code, can be written as $\hat{\rho} = \mathcal{F}|\Psi_\nu\rangle\langle\Psi_\nu| + (1 - \mathcal{F})\hat{I}/2^\nu$, where $\mathcal{F}$ is the fidelity of attaining the five-qubit code $\nu = 5$ (see Fig. 4.8a)
or the Steane code $\nu = 7$ (see Fig. 4.8b). In the figure, we remark that there are two
different probabilities- single-qubit probability p_1 and two-qubit probability p_2: p_1 is
used in $|+\rangle$ state preparation for individual qubit while p_2 involves in the two-qubit
gate error.

---

[3] As one sees that the number of qubits and the number two-qubit gates needed for both the five-
qubit and the Steane codes are in the similar order, one expects to see the similar result for the
Steane code, Fig. 4.8b, as compared to the five-qubit one as seen in Fig. 4.8a. The main reason of
the difference is that we have assumed the projective measurement error $p_m = 0.01$ for the three
orange colored qubits.

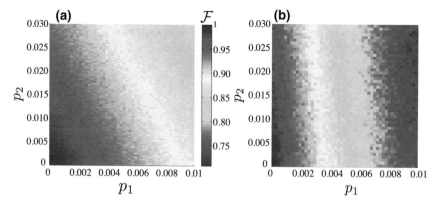

Fig. 4.8 Monte Carlo simulation results. **a** Fidelities of the five-qubit code, where the average fidelity value is taken over 5000 runs and **b** the Steane code, where the average fidelity value is taken over 1000 runs, are plotted against single-qubit gate error probability p_1 and two-qubit gate error probability p_2

4.3 Summary and Discussions

To sum up, we have proposed a possible realization of the five-qubit quantum error correction code and the Steane code in an array of superconducting flux qubits galvanically coupled to a coplanar waveguide resonator that mediates ultrafast two-qubit interactions. The system operates in the USC regime, in which two-qubit gates of subnanosecond timescale are numerically demonstrated. At this timescale, it is strenuous for the gate errors to be limited by the coherence time of the qubit and the resonator in the galvanic configuration [14, 15], whose rough estimation is 10–100 ns and 160–500 ns, respectively [16]. However, recent randomized benchmarking techniques in circuit QED technologies [17, 18] have shown that the error per gate can be reduced to about 0.5%. This precedent might encourage the realization of our approach, in which fidelities in excess of 75% could be achieved. Also, imperfect measurements are significant sources of errors in the construction of cluster states. However, extremely fast measurements with 99% fidelity have been demonstrated in Ref. [13]. Moreover, in light of current developments of large microwave cavity arrays, and following ideas from freely scalable quantum technologies developed in Ref. [19], one may think of scaling up our system to a two-dimensional array with nearest-neighbor coupling between cavities mediated by superconducting quantum interference devices. As already established in [19], the scaling up to large architectures does not imply increasing the number of physical qubits inside a unique device; instead, it has been proven that linking cells to one another via noisy channels is fault tolerant if entanglement purification is performed with high fidelity. Thus, we believe our proposal, with all the advanced technologies in the superconducting circuits, might pave a promising avenue for implementing large-scale QECCs or topological codes [20–24] in ultrafast timescale. This work has led to the publication

[25]: T. H. Kyaw, D. Herrera-Martí, E. Solano, G. Romero and L.-C. Kwek, *Creation of quantum error correcting codes in the ultrastrong coupling regime, Phys. Rev. B* **91**, 064503 (2015). http://journals.aps.org/prb/abstract/10.1103/PhysRevB. 91.064503.

References

1. Nielsen MA, Chuang IL (2000) Quantum computation and quantum information. Cambridge University Press
2. Steane AM (1996) Error correcting codes in quantum theory. Phys Rev Lett 77:793
3. Raussendorf R, Briegel HJ (2001) A one-way quantum computer. Phys Rev Lett 86:5188
4. Ozeri R (2013) Heisenberg limited metrology using quantum error-correction codes. arXiv:1310.3432
5. Arrad G, Vinkler Y, Aharonov D, Retzker A (2014) Increasing sensing resolution with error correction. Phys Rev Lett 112:150801
6. Bourassa J, Gambetta JM, Abdumalikov A, Astafiev O, Nakamura Y, Blais A (2009) Ultrastrong coupling regime of cavity QED with phase-biased flux qubits. Phys Rev A 80:032109
7. Leib M, Deppe F, Marx A, Gross R, Hartmann MJ (2012) Networks of nonlinear superconducting transmission line resonators. New J Phys 14:075024
8. Leib M, Hartmann MJ (2014) Synchronized switching in a Josephson junction crystal. Phys Rev Lett 112:223603
9. Romero G, Ballester D, Wang YM, Scarani V, Solano E (2012) Ultrafast quantum gates in circuit QED. Phys Rev Lett 108:120501
10. Jaynes ET, Cummings FW (1963) Comparison of quantum and semiclassical radiation theories with application to the beam maser. Proc IEEE 51:89
11. Wang YD, Chesi S, Loss D, Bruder C (2010) One-step multiqubit Greenberger-Horne-Zeilinger state generation in a circuit QED system. Phys Rev B 81:104524
12. Herrera-Martí DA, Rudolph T (2013) Loss tolerance with a concatenated graph state. Quant Inf Comp 13:0995
13. Jeffrey E, Sank D, Mutus JY, White TC, Kelly J, Barends R, Chen Y, Chen Z, Chiaro B, Dunsworth A et al (2014) Fast accurate state measurement with superconducting qubits. Phys Rev Lett 112(19):190504
14. Niemczyk T, Deppe F, Huebl H, Menzel EP, Hocke F, Schwarz MJ, Garcia-Ripoll JJ, Zueco D, Hummer T, Solano E, Marx A, Gross R (2010) Circuit quantum electrodynamics in the ultrastrong-coupling regime. Nat Phys 6:772
15. Forn-Díaz P, Lisenfeld J, Marcos D, García-Ripoll JJ, Solano E, Harmans CJPM, Mooij JE (2010) Observation of the Bloch-Siegert shift in a qubit-oscillator system in the ultrastrong coupling regime. Phys Rev Lett 105:237001
16. Forn-Díaz P at Institute for Quantum Computing University of Waterloo (private communication)
17. Chow JM, Gambetta JM, Magesan E, Abraham DW, Cross AW, Johnson BR, Masluk NA, Ryan CA, Smolin JA, Srinivasan SJ et al (2014) Implementing a strand of a scalable fault-tolerant quantum computing fabric. Nat Comm 5:4015
18. Barends R, Kelly J, Megrant A, Veitia A, Sank D, Jeffrey E, White TC, Mutus J, Fowler AG, Campbell B et al (2014) Superconducting quantum circuits at the surface code threshold for fault tolerance. Nature 508(7497):500
19. Nickerson NH, Fitzsimons JF, Benjamin SC (2014) Freely scalable quantum technologies using cells of 5-to-50 qubits with very lossy and noisy photonic links. Phys Rev X 4:041041
20. Dennis E, Landahl A, Kitaev A, Preskill J (2002) Topological quantum memory. J Math Phys 43:4452

21. Raussendorf R, Harrington J, Goyal K (2006) A fault-tolerant one-way quantum computer. Ann Phys 321:2242
22. Raussendorf R, Harrington J (2007) Fault-tolerant quantum computation with high threshold in two dimensions. Phys Rev Lett 98:190504
23. Barrett SD, Stace TM (2010) Fault tolerant quantum computation with very high threshold for loss errors. Phys Rev Lett 105:200502
24. Wang DS, Fowler AG, Hollenberg LCL (2011) Surface code quantum computing with error rates over 1. Phys Rev A 83:020302(R)
25. Kyaw TH, Herrera-Martí DA, Solano E, Romero G, Kwek L-C (2015) Creation of quantum error correcting codes in the ultrastrong coupling regime. Phys Rev B 91:064503

Chapter 5
Quantum Memory in the USC Regime

Creativity is intelligence having fun.

—Albert Einstein

In this chapter, we envisage a quantum memory cell composed of a superconducting flux qubit galvanically coupled [1] to a microwave resonator in the circuit QED framework. With the proposed cell, we would like to scale-up the entire structure into a three dimensional setup (see Chap. 7). We require that the input states are in the form of flying microwave photons [2–5], and study the storage and retrieval of single- and two-qubit quantum states. We find that these processes can be carried out with good fidelity even with the presence of noise. In addition, due to the recent technological advancements in superconducting circuits, pushed forward by IBM [6, 7] and Google [8, 9], we are convinced that our proposal will be put in good use one day by experimentalists. We believe our proposal might pave a way toward scalable quantum random access memory (QRAM) [10, 11] and distributed quantum interconnects [12, 13], which in turn might steer the whole field toward novel applications ranging from entangled state cryptography [14, 15], teleportation [16], purification [17, 18], fault-tolerant quantum computation [19] to quantum simulations. Lastly, with the physical implementation of QRAM within the superconducting circuits, our proposal is a step forward to experimentally realize many quantum machine learning algorithms based on the HHL[1] algorithm [20], Ref. [21] (Nature physics News & View) and references therein.

---

[1]HHL is named after the three researchers: Aram Harrow, Avinatan Hassidim, and Seth Lloyd.

© Springer Nature Switzerland AG 2019
T. H. Kyaw, *Towards a Scalable Quantum Computing Platform in the Ultrastrong Coupling Regime*, Springer Theses, https://doi.org/10.1007/978-3-030-19658-5_5

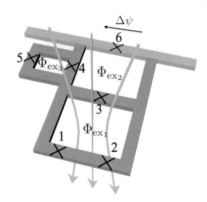

5.1   Quantum Memory Cell

The basic element of our quantum memory proposal is shown in Fig. 5.1, whose
Hamiltonian has been derived and discussed in the previous Chap. 4. It is

$$\hat{H} = \frac{\omega_{eg}}{2}\hat{\sigma}_z + \omega_{\mathrm{cav}}\hat{a}^\dagger\hat{a} + \hat{H}_{\mathrm{int}} \tag{5.1}$$

with an effective tunable interaction Hamiltonian

$$\hat{H}_{\mathrm{int}} = -2E_J\beta\cos\left(\pi f_{\mathrm{ex}_3}\right)\sum_{n=1,2}(\Delta\psi)^n\sum_{\mu=x,y,z}c_\mu^{(n)}\hat{\sigma}_\mu, \tag{5.2}$$

where E_J is the Josephson energy, β is a parameter that depends on the Josephson
junctions size, $\Phi_0 = h/2e$ is the flux quantum, and Φ_{ex_3} is an external flux through
a superconducting loop. The latter, in turn, allows one to switch on/off the qubit–
resonator coupling strength. $\Delta\psi$ stands for the phase slip shared by the resonator
and the flux qubit. And, the coefficients $c_\mu^{(n)}$'s can be tuned [22–24] at will via
additional external fluxes (c.f. Chap. 4). Moreover, the circuit exhibits a Z_2 parity
symmetry (see Chap. 3) and its dynamics is governed by the quantum Rabi [25] model
(c.f. Eq. 5.1)

$$\hat{H}_{\mathrm{Rabi}} = \frac{\omega_{eg}}{2}\hat{\sigma}_z + \omega_{\mathrm{cav}}\hat{a}^\dagger\hat{a} + g(t)\hat{\sigma}_x(\hat{a} + \hat{a}^\dagger), \tag{5.3}$$

where ω_{eg}, ω_{cav}, and g stand for the qubit frequency, cavity frequency, and tunable
qubit–resonator coupling strength, respectively. In addition, $\hat{a}(\hat{a}^\dagger)$ is the bosonic
annihilation(creation) operator, and $\hat{\sigma}_{x,z}$ are the Pauli matrices acting on the qubit. A
compelling feature of Hamiltonian (5.3) is that for ratios $g/\omega_{\mathrm{cav}} \gtrsim 0.8$, the ground
and first excited states can be approximated [26] as

$$|\psi_G\rangle \simeq \frac{1}{\sqrt{2}}(|+\rangle|-\alpha\rangle - |-\rangle|\alpha\rangle),$$

$$|\psi_E\rangle \simeq \frac{1}{\sqrt{2}}(|+\rangle|-\alpha\rangle + |-\rangle|\alpha\rangle), \tag{5.4}$$

where $|\alpha\rangle$ is a coherent state for the resonator field with amplitude $|\alpha| = g/\omega_{\text{cav}}$, and $|\pm\rangle = (|\uparrow\rangle \pm |\downarrow\rangle)/\sqrt{2}$ are the eigenstates of $\hat{\sigma}_x$. $\downarrow$ ($\uparrow$) stands for the ground (excited) state of the two-level system. The states $|\psi_{G/E}\rangle$ form a robust parity-protected qubit [26] whose coherence time can be up to $\tau_{\text{coh}} \gtrsim 10^5/\omega_{eg}$.

In the following, let us outline why the light–matter coupling system at the USC limit can become a quantum memory device as claimed by Ref. [26]. The first evidence comes from Chap. 3 where we have mentioned that the USC quantum states satisfy $\mathbb{Z}_2$ parity symmetry. In particular, the states $|\psi_{G/E}\rangle$ belong to two different parity eigenvalue $+1/-1$. Hence, the transition from $|\psi_E\rangle \rightarrow |\psi_G\rangle$ is forbidden unless there is any parity broken interaction present in the USC system. Second, let us try to understand in terms of an open quantum system analysis detailed in Appendix B, where we have the microscopic master equation for a USC system weakly coupled to some environment

$$\dot{\rho}_S(t) = \sum_{j,k>j} i\omega_{kj}\langle j|\hat{\rho}_S(t)|k\rangle|j\rangle\langle k| \tag{5.5}$$

$$+ \sum_{\alpha} \sum_{j,k>j} \Gamma_\alpha^{jk}\left(|j\rangle\langle k|\hat{\rho}_S(t)|k\rangle\langle j| - \frac{1}{2}\left(|k\rangle\langle k|\hat{\rho}_S(t) + \hat{\rho}_S(t)|k\rangle\langle k|\right)\right),$$

And, the decay rates Γs are proportional to $|C_{jk}|^2$ where $C_{jk} = -i\langle j|\hat{A}_\alpha|k\rangle|$, $\hat{A}_\alpha = (\hat{a} + \hat{a}^\dagger)$, $\hat{\sigma}_x$, $\hat{\sigma}_y$ and $\hat{\sigma}_z$. $|j/k\rangle$ are eigenstates of the quantum Rabi model, Eq. (5.3). In accordance to Ref. [26], $|\psi_{G/E}\rangle$ are robust with respect to a local and static perturbation $\hat{H}_{y,z}^{\text{pert}} = h_y\hat{\sigma}_y + h_z\hat{\sigma}_z$, where h_y and h_z are some random perturbation amplitudes. In this particular case, we have

$$|C_{10}|^2 = |\langle\psi_G|\hat{\sigma}_{y/z}|\psi_E\rangle|^2 \propto |\langle-\alpha|\alpha\rangle|^2 = \exp(-4|\alpha|^2) \sim \exp\left(-4\frac{g^2}{\omega_{\text{cav}}^2}\right). \tag{5.6}$$

That means suppression of the local perturbation $\hat{H}_{y,z}^{\text{pert}}$ is readily available in the USC limit. And the error suppression gets better as the light–matter coupling increases from the strong coupling all the way to the ultrastrong coupling. In the following, we are mainly interested in the circuit QED setup in the presence of $\hat{H}_{y,z}^{\text{pert}}$.

The third observation comes from the quantum Rabi model spectrum seen in Fig. 5.2. On the left, near $g \sim 0$, we are in the weak coupling regime, where the Jaynes–Cummings Hamiltonian plays an important role. Hence, the ground and first excited states with very little light–matter interaction are: specifically $|\psi_0\rangle = |\downarrow, 0_F\rangle$ & $|\psi_1\rangle = |\uparrow, 0_F\rangle$. Here, $|0_F\rangle$ represent zero photon/vacuum inside the resonator, in the Fock state representation. When $g/\omega_{\text{cav}} \geq 0.1$, one is in the USC limit. The ground and first excited states are of the form $|\psi_{G/E}\rangle$, Eq. (5.4), as $g/\omega_{\text{cav}} \rightarrow 1$.

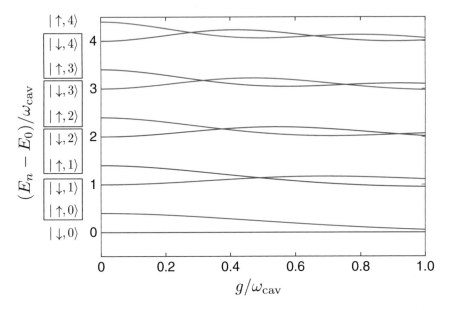

Fig. 5.2 Spectrum of the quantum Rabi model, Eq. (5.3), for $\omega_{eg} = 0.4$, $\omega_{\text{cav}} = 1$, and $0 \le g/\omega_{\text{cav}} \le 1$ in the spaces with positive (red lines) and negative (blue lines) parity. On the left side, we have states with the light–matter interaction close to 0. The states within a rectangular box are the usual Jaynes–Cummings doublets

In between the weak and USC couplings, the two states do not cross. Therefore, one can invoke the adiabatic theorem to tune the light–matter coupling such that an instantaneous eigenstate remains in the same eigenstate as the parameter g is being tuned adiabatically. The idea is that we like to perform any necessary quantum computing in the weak coupling limit. After the computations, we store the desired quantum information in the USC limit by adiabatically tuning up the light–matter coupling. That forms the basics of our quantum memory cell.

5.2   Generating and Catching Flying Qubits

Single photons propagating through linear devices are well suited as information carriers because they possess long coherence length and can be encoded with useful information. In our case, a flying microwave photon is generated from circuit QED platforms [2–5, 27–29], and a qubit is encoded in a linear superposition of zero ($|0\rangle$) and one photon ($|1\rangle$) Fock states. Recently, Ref. [5] shows that if a photonic wave packet emitted from a source has a temporally symmetric profile, it overcomes the impedance mismatch problem when a flying qubit impinges onto a resonator. With all these advancements in superconducting circuit QED technologies, we envision our memory cell be located on the pathway of a single microwave photon to accomplish quantum information storage as shown in Fig. 5.3.

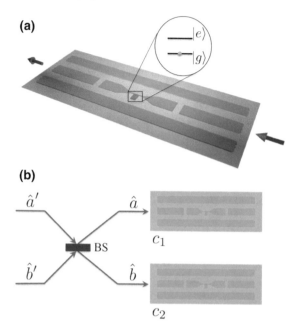

Fig. 5.3 Schematic of circuit QED design for storage and retrieval of an unknown single- and two-qubit states. **a** A USC memory cell element, composed of a qubit–resonator system operating at the USC regime. **b** Two flying microwave photons, with modes $\hat{a}'$ and $\hat{b}'$, come in and pass through a beam splitter (BS) implemented by a superconducting quantum interference device (SQUID) to form a two-qubit entangled state, which is then stored in two USC qubits located at a distance apart

5.3 Storage and Retrieval of Flying Qubits

In the following section, we outline a protocol that allows storage and retrieval of quantum information to and from the fundamental quantum memory cell we have described above. It is achieved by adiabatically tuning the qubit–resonator coupling strength, from the Jaynes–Cummings (JC) to USC regime. In particular, we propose a USC memory cell element (see Fig. 5.3a) that can be designed by the flux qubit architecture, which provides a tunable qubit–resonator coupling. The latter can be implemented by using a superconducting quantum interference device (SQUID) as proposed for qubit–qubit coupling in Refs. [23, 24].

5.3.1 The Storage Protocol

The storage of quantum information into our USC memory cell is done through three steps. First, we cool down the entire system to its ground state in the USC limit. Second, the qubit frequency is tuned to be off-resonant with the resonator frequency, i.e., $\omega_{\text{cav}} > \omega_{eg}$, while the qubit–resonator coupling strength g is adiabatically tuned toward the strong coupling regime where $g/\omega_{\text{cav}} \ll 1$, where the coupling is much larger than any decoherence rate in the system. In this regime, the ground and first excited states of the qubit–resonator system are $|\psi_0\rangle = |g, 0\rangle$ and $|\psi_1\rangle = |e, 0\rangle$, respectively. Since we have adiabatically tuned the coupling from the USC to the strong coupling regime, our initial USC ground state is then mapped to the JC ground state. At this stage, our memory cell is ready for information storage. When a flying qubit with an unknown quantum state, for instance, $|\Psi\rangle_F = \alpha_F |0\rangle_F + \beta_F |1\rangle_F$ comes in contact with the memory cell as shown in Fig. 5.3a, the encoded flying qubit is transferred to the flux qubit due to the JC interaction. The subscript F stands for the flying qubit. Therefore, state of the memory cell becomes $|\psi_s\rangle = (\alpha_F |\downarrow\rangle + \beta_F |\uparrow\rangle) \otimes |0\rangle$. At last, we turn on the qubit–resonator coupling adiabatically toward the USC regime. For simplicity, we consider a linear adiabatic switching scheme such that $g(t) = (\cos(\pi f_3) - \Delta \pi f_3 \sin(\pi f_3)t/T)g_0$, with T total evolution time and $f_3 = f_{\text{ex}_3}$. The same linear ramp in Chap. 4 is applied here. In Fig. 5.4a, b, we show the storage and retrieval processes for a quantum state $|\psi_s\rangle = \alpha_F |\psi_0\rangle + \beta_F |\psi_1\rangle$, and Fig. 5.4c, d show the ground state $|\psi_0\rangle$ and the first excited state $|\psi_1\rangle$ adiabatically follow the instantaneous eigenstates such that $|\psi_0\rangle \rightarrow |\psi_G\rangle$ and $|\psi_1\rangle \rightarrow |\psi_E\rangle$. In this manner, we can encode important information onto the parity-protected qubit basis.

5.3.2 The Retrieval Protocol

Retrieval (decoding) process is reverse of the storage process and is achieved by adiabatically switching off the qubit–resonator coupling strength (g) from the USC to strong coupling regime such that $|\psi_G\rangle \rightarrow |\psi_0\rangle$ and $|\psi_E\rangle \rightarrow |\psi_1\rangle$.

We note that the time for storage and retrieval of quantum information is several order of magnitude faster than the coherence time of the parity-protected qubit, which is about $T_{\text{coh}} \sim 40\,\mu\text{s}$ for a coupling strength $g_0/\omega_{eg} \sim 1.5$ [26]. For instance, if we consider a flux qubit with energy $\omega_{eg}/2\pi \sim 2$ GHz, and a cavity of frequency $\omega_{\text{cav}}/2\pi \sim 5$ GHz, our system reaches the USC regime with $g_0/\omega_{\text{cav}} = 0.6$. For the linear adiabatic switching scheme with the above parameters, we estimate total time for storage/retrieval of a qubit is about $\tilde{T} \approx 2$–8 ns.

5.3.3 Unavoidable Phase Imprinting During Storage and Retrieval

At the end of an adiabatic evolution, the state $|\tilde{\psi}\rangle = \alpha_F |\psi_G\rangle + \beta_F |\psi_E\rangle$ is desired. However, the state after the evolution might become $|\tilde{\psi}(T)\rangle = \alpha_F |\psi_G(T)\rangle + \beta_F e^{i\theta(T)} |\psi_E(T)\rangle$, with a relative phase $\theta(T)$ resulting from the dynamical and geometrical

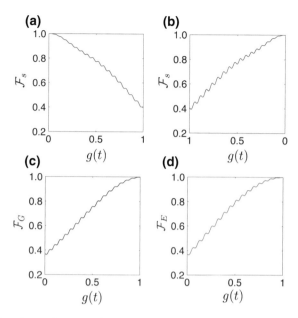

Fig. 5.4 Fidelity plots. **a** Storage and **b** retrieval processes for a quantum state $|\psi_s\rangle = \alpha_F|\psi_0\rangle + \beta_F|\psi_1\rangle$. In both cases, we plot the fidelity between the initial $|\psi_s\rangle$ and the instantaneous state $|\psi(t)\rangle$, i.e., $\mathcal{F}_s = |\langle\psi_s|\psi(t)\rangle|^2$. Any arbitrary state $|\psi\rangle = u|\psi_0\rangle + v|\psi_1\rangle$ can be stored and retrieved with unit fidelity. **c** Fidelity between the approximated ground state in Eq. (5.4) and the instantaneous ground state $\mathcal{F}_G = |\langle\psi_G|\psi_G(t)\rangle|^2$. **d** Fidelity between the approximated first excited state in Eq. (5.4) and the instantaneous first excited state $\mathcal{F}_E = |\langle\psi_E|\psi_E(t)\rangle|^2$. For all the simulations, we choose the system parameters as $\omega_{\text{cav}} = 1$, $\omega_{eg} = 0.1 \, \omega_{\text{cav}}$, $g_0 = \omega_{\text{cav}}$, and the total evolution $T = 105/\omega_{\text{cav}}$

effects [30]. Hence, we need to keep track of a relative phase during the storage and retrieval processes. In order to figure out which phase $\theta(t)$ optimizes the processes, in Fig. 5.5a, b, we plot the fidelity $\mathcal{F}(g, \theta) = |\langle\tilde{\psi}|\psi(t)\rangle|^2$ between the state $|\tilde{\psi}\rangle = \alpha_F|\psi_G\rangle + \beta_F e^{i\theta}|\psi_E\rangle$ and the state $|\psi(t)\rangle$, which has adiabatically evolved from the initial state $|\psi_s\rangle = \alpha_F|\psi_0\rangle + \beta_F|\psi_1\rangle$. In these simulations, we find the fidelity over the landscape of $\theta \in [0, 2\pi]$ versus the qubit–resonator coupling strength $g(t)$, for two different total evolution time $T = 105/\omega_{\text{cav}}$ (see Fig. 5.5a), and $T = 120/\omega_{\text{cav}}$ (see Fig. 5.5b). White lines show the phase θ_{opt}, which optimizes the fidelity $\mathcal{F}$ for both cases. Notice that the maximum fidelity and the optimal phase θ depend strongly on the system parameters and the total evolution time T. Thus, we require, for each USC memory cell, to find out the parameter T that maximizes the fidelity only once. When T is known, the cell can always be operated at that specific parameter for storing and retrieving unknown quantum states. Therefore, the time T might be a benchmark to characterize our potential USC quantum memory devices, in the same way as hard disk drives of the classical computer are being characterized by their seek time and latency.

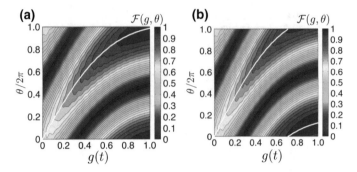

Fig. 5.5 Contour plots of the fidelity $\mathcal{F} = |\langle \tilde{\psi} | \psi(t) \rangle|^2$ between the state $|\tilde{\psi}\rangle = \alpha_F |\psi_G\rangle + \beta_F e^{i\theta} |\psi_E\rangle$ and the state $|\psi(t)\rangle$, which has adiabatically evolved from the initial state $|\psi_s\rangle = \alpha_F |\psi_0\rangle + \beta_F |\psi_1\rangle$. **a** The total evolution time is set to $T = 105/\omega_{\text{cav}}$. **b** The evolution time is $T = 120/\omega_{\text{cav}}$. The white colored lines stand for the phase which maximized the fidelity $\mathcal{F}$. In these simulations, the parameters are $\omega_{\text{cav}} = 1$, $\omega_{eg} = 0.1 \, \omega_{\text{cav}}$, and $g_0 = \omega_{\text{cav}}$

5.3.4   Storage and Retrieval of Entangled Flying Qubits

Additionally, storage and retrieval of entangled states in two separate USC cells is feasible. To demonstrate such a process, we let two bosonic fields to interact via the SQUID, simulating a Hong-Ou-Mandel setup [31] as shown in Fig. 5.3b. Let us suppose that we have an initial state $|\psi_0\rangle = |0\rangle_{\hat{a}'} |1\rangle_{\hat{b}'}$. After experiencing a beam splitter interaction, we have two-photon entangled state $|\psi_0'\rangle = \frac{1}{\sqrt{2}}(|0\rangle_{\hat{a}}|1\rangle_{\hat{b}} + |1\rangle_{\hat{a}}|0\rangle_{\hat{b}})$, which enters two cavities c_1 and c_2, each containing a flux qubit prepared in its ground state. This process allows the cavities to be prepared in the state $|\overline{\Psi}_0\rangle = \frac{1}{\sqrt{2}}(|0\rangle_{c_1}|1\rangle_{c_2} + |1\rangle_{c_1}|0\rangle_{c_2}) \otimes |gg\rangle$. Following the same procedure, we tune the qubits toward resonance with its respective cavity such that we arrive at the state $|\Psi_0\rangle = \frac{1}{\sqrt{2}}(|\downarrow\uparrow\rangle + |\uparrow\downarrow\rangle) \otimes |00\rangle_{c_1 c_2}$. With our protocol above, the state is eventually mapped to a parity-protected state $|\tilde{\Psi}_0\rangle = \frac{1}{\sqrt{2}}(|\psi_G\rangle|\psi_E\rangle + |\psi_E\rangle|\psi_G\rangle)$. In Fig. 5.6a, b, we show the numerical simulations for the storage and retrieval processes of the entangled state $|\Psi_0\rangle$.

5.3.5   Open Quantum System Treatment

It is inevitable that any practical and realistic quantum system operates in noisy environment and so is our memory element. Moreover, it is well known that the standard quantum optics master equation technique is not valid for any value of the qubit–field coupling [32]. Hence, we follow the microscopic derivation [33] and the open system analysis of a USC system [34]. Then, we obtain the master equation (see Appendix B)

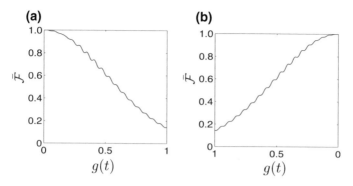

Fig. 5.6 **a** Storage process for an entangled state $|\Psi_0\rangle = \frac{1}{\sqrt{2}}(|ge\rangle + |eg\rangle) \otimes |00\rangle_{c_1 c_2}$. **b** Retrieval process. In both cases, we plot the fidelity between the initial state $|\Psi_0\rangle$, and the instantaneous state $|\psi(t)\rangle$, $\bar{\mathcal{F}} = |\langle\Psi_0|\psi(t)\rangle|^2$. In the simulations, we have chosen $\omega_{cav} = 1$, $\omega_{eg} = 0.1\,\omega_{cav}$, $g_0 = \omega_{cav}$, and the evolution time $T = 105/\omega_{cav}$

$$\dot{\hat{\rho}}(t) = i[\hat{\rho}(t), \hat{H}_S] + \mathcal{L}_a\hat{\rho}(t) + \mathcal{L}_\sigma\hat{\rho}^\sigma(t), \tag{5.7}$$

where $\hat{H}_S = \hat{H}_{Rabi}$ of Eq. (5.1) and $\sigma = x, y, z$. $\mathcal{L}_a$ and $\mathcal{L}_\sigma$ are Liouvillian superoperators with

$$\mathcal{L}_\nu\hat{\rho}(t) = \sum_{j,k>j} \Gamma_\nu^{jk}(1 + \bar{n}_\nu(\Delta_{kj}, T))\mathcal{D}[|j\rangle\langle k|]\hat{\rho}(t)$$

$$+ \sum_{j,k>j} \Gamma_\nu^{jk}\bar{n}_c(\Delta_{kj}, T)\mathcal{D}[|k\rangle\langle j|]\hat{\rho}(t), \tag{5.8}$$

where $\nu = a, x, y, z$, $\mathcal{D}[\hat{O}]\hat{\rho} = (2\hat{O}\hat{\rho}\hat{O}^\dagger - \hat{\rho}\hat{O}^\dagger\hat{O} - \hat{O}^\dagger\hat{O}\hat{\rho})/2$, T is the temperature of the thermal bath and $\bar{n}_\nu(\Delta_{kj}, T))$ is the number of thermal photons feeding the system from all the possible $|k\rangle \rightarrow |j\rangle$ transitions. Here, states $|j\rangle$ are eigenstates of $\hat{H}_{Rabi}$ with respective eigenenergy $\hbar\omega_j$, i.e., $\hat{H}_S|j\rangle = \hbar\omega_j|j\rangle$. To arrive at the numerical simulation shown in Fig. 5.7, we assume our system is in a very low temperature environment, i.e., $T \simeq 0$, and the relaxation coefficients take the form [34] $\Gamma_\nu^{jk} = \gamma_\nu \frac{\Delta_{kj}}{\omega_0}|C_{jk}^\nu|^2$, where γ_ν are standard weak coupling damping rates, $\Delta_{kj} = \omega_k - \omega_j$, and $C_{jk}^\nu = -i\langle j|\hat{\Theta}|k\rangle$, $(\hat{\Theta} = \hat{a} - \hat{a}^\dagger)$ for $\nu = a$ and $(\hat{\Theta} = \hat{\sigma}_{x,y,z})$ for $\nu = \{x, y, z\}$, respectively. We show numerical results for the storage and retrieval processes of an arbitrary superposed state $|\psi_s\rangle$ in presence of external noises, in Fig. 5.7. With our scheme and a simple decoherence model, we estimate fidelity of $\mathcal{F}_s = 0.9939$ at the end of the retrieval process.

Lastly, we like to mention that the decoherent rates Γ_α used in our numerical simulations in Fig. 5.7 are taken from one particular value of g/ω_{cav} in the USC limit [34], while we are tuning the light–matter coupling adiabatically. The main assumption here is that the various decay rates throughout the adiabatic evolution remain constant, which might not be the case in the actual experimental setup. Hence,

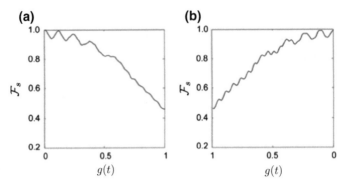

Fig. 5.7 Fidelity plots. **a** Storage process for a quantum state $|\psi_s\rangle = \alpha_F|\psi_0\rangle + \beta_F|\psi_1\rangle$. **b** Retrieval process. In both cases, we plot the fidelity between the initial $|\psi\rangle_s$ and the instantaneous state $|\psi(t)\rangle$, i.e., $\mathcal{F}_s = |\langle\psi_s|\psi(t)\rangle|^2$, while we introduce external noises $\Gamma_x = \Gamma_y = \Gamma_z = 10^{-3}\omega_{eg}$ and $\Gamma_r = 10^{-4}\omega_{eg}$ into our close system with the help of the microscopic derivation [34]. Here, $\Gamma_x, \Gamma_y, \Gamma_z$ and Γ_r are rates of bit-flip noises, dephasing noise of the qubit, and the resonator relaxation, respectively. The system parameters are $\omega_{cav} = 1$, $\omega_{eg} = 0.1\ \omega_{cav}$, $\Omega_0 = \omega_{cav}$, and the total evolution time is $T = 105/\omega_{cav}$

we need better modeling techniques to actually capture the correct physics, which we would leave it for the future work.

5.4   Scaling up to Two Dimensions

5.4.1   Cavity Network

In order to transfer the state of a qubit along a given path in a cavity network, interactions between neighboring resonators must be selectively turned on and off on demand. This can be done via connecting more resonators to a single circuit node, and then grounding that node through a superconducting quantum interference device (SQUID), as shown in Fig. 5.8. A SQUID is a superconducting loop interrupted by two JJs. When a SQUID is perfectly symmetrical, it behaves as a single JJ. Under the condition of same loop inductance, a perfect symmetric SQUID generates the largest E_J modulation. Its effective Josephson energy can be tuned by threading the loop with an external magnetic flux (see Chap. 2). Indeed, this device can be seen as a tunable inductance, shunted by a small capacitance. The SQUID inductance can be written as $L_J = \left(\frac{\Phi_0}{2\pi}\right)^2 \frac{1}{E_J \cos\left(\frac{\pi\Phi_{ext}}{\Phi_0}\right)}$.

If the system parameters are chosen in order to make the SQUID impedance much smaller than that of the resonators, the electrical potential at the node can be approximated to zero. Such condition is naturally satisfied in most circuit QED experiments involving SQUIDs in the non-dissipative regime. This enables us to define well

Fig. 5.8 A node of the cavity network, showing four single-mode resonators grounded through a SQUID device. The color scheme suggests that the frequencies ω_i of the four cavities are different so that direct interactions among them are off-resonance. Hopping interactions between any cavity pair can be activated by driving the SQUID with an external flux, which must be oscillating with the frequency given by sum of the two-cavity bare frequencies. The light blue square in each cavity represents a tunable flux qubit, together with the cavity forming a USC quantum memory

separated spatial modes for the electromagnetic field in the different resonators. A direct coupling between resonators is then given by the inductive energy term of the SQUID. The strength of this interaction depends on the SQUID impedance and the external flux threading the device. After quantization, the Hamiltonian describing a single network node can be written as (see Ref. [35] for detailed derivation)

$$\hat{\mathcal{H}} = \sum_l \omega_l \hat{a}_l^\dagger \hat{a}_l - \sum_{l,r} \alpha_{l,r}(t) \left(\hat{a}_l^\dagger + \hat{a}_l\right)\left(\hat{a}_r^\dagger + \hat{a}_r\right), \qquad (5.9)$$

where ω_l is frequency of the lth resonator and the indexes l, r run over all resonators pairs. The corresponding coupling parameters are given by

$$\alpha_{l,r}(t) = \frac{\varphi_0^2}{E_J(\phi_{\text{ext}})} \sqrt{\frac{\omega_l \omega_r}{C_l C_r}} \frac{1}{Z_l Z_r}, \qquad (5.10)$$

where C_i and Z_i indicate the resonator effective capacitance and impedance, respectively. When the resonators are off-resonance, i.e., when $|\omega_l - \omega_r| \gg \alpha_{l,r}$, their interaction is negligible, as far as the coupling strength is constant. Nevertheless, the SQUID can be driven by an external magnetic flux oscillating at frequency comparable with that of the resonators [36]. In this way, the direct interaction term between any resonator pair can be tuned on resonance when the corresponding frequency matching conditions are satisfied. In particular, we can set $\pi\Phi_{\text{ext}}/\Phi_0 = \bar{\phi} + \Delta\cos(\omega_d t)$, where $\bar{\phi}$ is a constant offset and Δ is the amplitude of a small harmonic drive, whose frequency is given by ω_d. Neglecting terms of the order of Δ^2, we obtain

$$\frac{1}{E_J\left(\Phi_{\text{ext}}\right)} \approx \frac{1}{\cos\bar{\phi}} + \frac{\sin\bar{\phi}}{\cos^2\bar{\phi}} \, \Delta \, \cos\left(\omega_d t\right). \tag{5.11}$$

To activate hopping interactions $(a_l^\dagger a_r + a_l a_r^\dagger)$ between two resonators, the driving frequency must satisfy the condition $\omega_d = |\omega_l - \omega_r|$. By controlling the external magnetic flux threading the SQUID device, this scheme allows to selectively turn on an inductive coupling between any two neighboring resonators, which would not interact otherwise.

5.5 Summary and Discussions

We have presented the basic tools for building a quantum memory based on a circuit QED architecture that operates in the USC regime of light–matter interaction. The storage/retrieval process for unknown quantum states, be single-qubit or two-qubit entangled states, can be accomplished by adiabatically switching on/off the qubit–resonator coupling strength. This work has resulted in one publication [37]: T. H. Kyaw, S. Felicetti, G. Romero, E. Solano and L.-C. Kwek, *Scalable quantum memory in the ultrastrong coupling regime, Sci. Rep.* **5**, 8621 (2015). http://www.nature.com/srep/2015/150302/srep08621/full/srep08621.html & one article in a proceedings [38]: T. H. Kyaw, S. Felicetti, G. Romero, E. Solano and L.-C. Kwek, $\mathbb{Z}_2$ *quantum memory implemented on circuit quantum electrodynamics, Proc. SPIE.* **9225**, 92250B (2014). http://spie.org/Publications/Proceedings/Paper/10.1117/12.2062893.

References

1. Bourassa J, Gambetta JM, Abdumalikov A, Astafiev O, Nakamura Y, Blais A (2009) Ultrastrong coupling regime of cavity QED with phase-biased flux qubits. Phys Rev A 80:032109
2. Wenner J, Yin Y, Chen Y, Barends R, Chiaro B, Jeffrey E, Kelly J, Megrant A, Mutus JY, Neill C, et al (2014) Catching time-reversed microwave coherent state photons with 99.4% absorption efficiency. Phys Rev Lett 112(21):21050
3. Houck AA, Schuster DI, Gambetta JM, Schreier JA, Johnson BR, Chow JM, Frunzio L, Majer J, Devoret MH, Girvin SM et al (2007) Generating single microwave photons in a circuit. Nature 449(7160):328
4. Yin Y, Chen Y, Sank D, O'Malley PJJ, White TC, Barends R, Kelly J, Lucero E, Mariantoni M, Megrant A et al (2013) Catch and release of microwave photon states. Phys Rev Lett 110(10):107001
5. Srinivasan SJ, Sundaresan NM, Sadri D, Liu Y, Gambetta JM, Yu T, Girvin SM, Houck AA (2014) Time-reversal symmetrization of spontaneous emission for quantum state transfer. Phys Rev A 89(3):033857

6. Chow JM, Gambetta JM, Magesan E, Abraham DW, Cross AW, Johnson BR, Masluk NA, Ryan CA, Smolin JA, Srinivasan SJ et al (2014) Implementing a strand of a scalable fault-tolerant quantum computing fabric. Nat Comm 5:4015

7. Chow JM, Gambetta JM, Rothwell MB, Rozen JR (2018) Modular array of vertically integrated superconducting qubit devices for scalable quantum computing. US Patent App. 15/871,443

8. Barends R, Kelly J, Megrant A, Veitia A, Sank D, Jeffrey E, White TC, Mutus J, Fowler AG, Campbell B et al (2014) Superconducting quantum circuits at the surface code threshold for fault tolerance. Nature 508(7497):500

9. Jeffrey E, Sank D, Mutus JY, White TC, Kelly J, Barends R, Chen Y, Chen Z, Chiaro B, Dunsworth A et al (2014) Fast accurate state measurement with superconducting qubits. Phys Rev Lett 112(19):190504

10. Giovannetti V, Lloyd S, Maccone L (2008a) Quantum random access memory. Phys Rev Lett 100:160501

11. Giovannetti V, Lloyd S, Maccone L (2008b) Architectures for a quantum random access memory. Phys Rev A 78:052310

12. Cirac JI, Ekert AK, Huelga SF, Macchiavello C (1999) Distributed quantum computation over noisy channels. Phys Rev A 59:4249

13. Kimble HJ (2008) The quantum internet. Nature 453:1023

14. Ekert AK (1991) Quantum cryptography based on Bell's theorem. Phys Rev Lett 67:661

15. Bennett CH (1995) Quantum information and computation. Phys Today 48:24

16. Bennett CH, Brassard G, Crépeau C, Jozsa R, Peres A, Wootters WK (1895) Teleporting an unknown quantum state via dual classical and Einstein-Podolsky-Rosen channels. Phys Rev Lett 70(13):1993

17. Bennett CH, Brassard G, Popescu S, Schumacher B, Smolin JA, Wootters WK (1996) Purification of noisy entanglement and faithful teleportation via noisy channels. Phys Rev Lett 76(5):722

18. Deutsch D, Ekert A, Jozsa R, Macchiavello C, Popescu S, Sanpera A (1996) Quantum privacy amplification and the security of quantum cryptography over noisy channels. Phys Rev Lett 77(13):2818

19. Nigg SE, Girvin SM (2013) Stabilizer quantum error correction toolbox for superconducting qubits. Phys Rev Lett 110:243604

20. Harrow AW, Hassidim A, Lloyd S (2009) Quantum algorithm for linear systems of equations. Phys Rev Lett 103(15):150502

21. Childs AM (2009) Quantum algorithms: equation solving by simulation. Nat Phys 5(12):861

22. Romero G, Ballester D, Wang YM, Scarani V, Solano E (2012) Ultrafast quantum gates in circuit QED. Phys Rev Lett 108:120501

23. Makhlin Y, Scöhn G, Shnirman A (1999) Josephson-junction qubits with controlled couplings. Nature 398:305

24. Grajcar M, Liu Y-X, Nori F, Zagoskin AM (2006) Switchable resonant coupling of flux qubits. Phys Rev B 74:172505

25. Braak D (2011) Integrability of the Rabi model. Phys Rev Lett 107:100401

26. Nataf P, Ciuti C (2011) Protected quantum computation with multiple resonators in ultrastrong coupling circuit QED. Phys Rev Lett 107:190402

27. Hofheinz M, Weig EM, Ansmann M, Bialczak RC, Lucero E, Neeley M, O'connell AD, Wang H, Martinis JM, Cleland AN (2008) Generation of Fock states in a superconducting quantum circuit. Nature 454(7202):310

28. Peng ZH, De Graaf SE, Tsai JS, Astafiev OV (2016) Tuneable on-demand single-photon source in the microwave range. Nat Comm 7:12588

29. Sathyamoorthy SR, Bengtsson A, Bens S, Simoen M, Delsing P, Johansson G (2016) Simple, robust, and on-demand generation of single and correlated photons. Phys Rev A 93(6):063823

30. Berry MV (1984) Quantal phase factors accompanying adiabatic changes. Pros R Soc Lond A 392:45

31. Lang C, Eichler C, Steffen L, Fink JM, Woolley MJ, Blais A, Wallraff A (2013) Correlations, indistinguishability and entanglement in hong-ou-mandel experiments at microwave frequencies. Nat Phys 9(6):345

32. Beaudoin F, Gambetta JM, Blais A (2011) Dissipation and ultrastrong coupling in circuit QED. Phys Rev A 84:043832
33. Breuer H-P, Petruccione F (2002) The theory of open quantum systems. Oxford University Press on Demand
34. Ridolfo A, Leib M, Savasta S, Hartmann MJ (2012) Photon blockade in the ultrastrong coupling regime. Phys Rev Lett 109:193602
35. Felicetti S, Sanz M, Lamata L, Romero G, Johansson G, Delsing P, Solano E (2014) Dynamical Casimir effect entangles artificial atoms. Phys Rev Lett 113(9):093602
36. Wilson CM, Johansson G, Pourkabirian A, Simoen M, Johansson JR, Duty T, Nori F, Delsing P (2011) Observation of the dynamical Casimir effect in a superconducting circuit. Nature 479(7373):376
37. Kyaw TH, Felicetti S, Romero G, Solano E, Kwek L-C (2015) Scalable quantum memory in the ultrastrong coupling regime. Sci Rep 5:8621
38. Kyaw TH, Felicetti S, Romero G, Solano E, Kwek L-C (2014) $\mathbb{Z}_2$ quantum memory implemented on circuit quantum electrodynamics. Proc SPIE 9225:92250B

Chapter 6
Catalytic Quantum Rabi Model

The chief enemy of creativity is good sense.

— Pablo Picasso

Apart from fundamental interests in light–matter interaction at the USC regime, the ultrastrong coupled qubit–cavity system or quantum Rabi system (QRS) is also extensively studied for its potential impetus to speed up quantum information processing at subnanosecond timescales [1–5], particularly within the framework of circuit quantum electrodynamics (QED) [6]. In this context, we have dedicated the previous two Chaps. 4 and 5 to use the tunable qubit–cavity coupling [7] to achieve the ultrafast quantum gates [4]. However, there is a drawback in those proposals, which is the need to use external magnetic fluxes threading many loops within a flux qubit configuration, during quantum gate operations. A standard qubit is in micrometer regime, therefore it is very hard to implement micrometer resolution magnetic fields without making any crosstalk between different superconducting loops. With the proposed framework presented in this chapter, the magnetic crosstalk problem could be overcome, and the setup can be realized experimentally.

Here, we will present a parity-preserving USC system, which we term as a catalytic quantum Rabi model, mediating effective interactions between two qubits. The proposal has four compelling characteristics that might have important implications in superconducting circuit-based quantum computing (from point (1) to (3)) and solid-state physics (point (4)) communities.

1. Strong two-qubit interaction with an increase in the qubit-cavity coupling strength of the QRS (g_p/ω_{cav}) is demonstrated.
2. A tunable qubit–qubit interaction could be performed by sweeping only the qubits energy gap for fixed QRS parameters, without needing complex flux qubit architectures [4, 7].

© Springer Nature Switzerland AG 2019
T. H. Kyaw, *Towards a Scalable Quantum Computing Platform in the Ultrastrong Coupling Regime*, Springer Theses,
https://doi.org/10.1007/978-3-030-19658-5_6

3. Manipulation of the qubits energy gap does not change the underlying $\mathbb{Z}_2$ symmetry, with which generalization to a system, with N qubits and a QRS, can easily be extended; thereby we provide an intuitive physical insight.
4. Enhanced excitation transfer between the two nonidentical qubits with increase in g_p/ω_{cav} is shown, while one qubit experiences an incoherent pumping and the other one experiences a loss mechanism. From extensive numerical studies, we provide an interesting physical insight that might shed some light on the cavity-enhanced exciton transport in disordered medium [8–10], especially within the context of polyatomic molecules in the USC regime [11].

6.1 The Simple Model

We consider a catalytic QRS (see Fig. 6.1), which is described by the quantum Rabi model [12, 13]

$$\hat{H}_p = \frac{\omega_p}{2}\hat{\sigma}_z^p + \omega_{cav}\hat{a}^\dagger\hat{a} + g_p\hat{\sigma}_x^p(\hat{a} + \hat{a}^\dagger). \tag{6.1}$$

Here, $\hat{a}(\hat{a}^\dagger)$ is the single mode bosonic annihilation(creation) operator and $\hat{\sigma}_{x,z}$ are the usual Pauli matrices. We denote ω_{cav} as the cavity mode frequency and ω_p and g_p as the qubit frequency and the qubit–cavity coupling strength, respectively. In addition, we have two additional qubits with frequencies ω_{q1} and ω_{q2}. They are coupled to the catalytic QRS via the cavity mode with coupling strengths g_1 and g_2. The Hamiltonian for our entire model reads

$$\hat{H} = \hat{H}_p + \sum_{n=1}^{N=2} \frac{\omega_{qn}}{2}\hat{\sigma}_z^n + g_n\hat{\sigma}_x^n(\hat{a} + \hat{a}^\dagger). \tag{6.2}$$

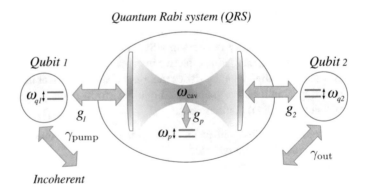

Quantum Rabi system (QRS)

Fig. 6.1 Our model: a qubit–cavity system interacting in the USC regime constitutes the QRS, while two additional qubits interact with the cavity mode. This system might also be considered as a building block for exciton transport mechanism, where qubit 1 is driven by an incoherent pumping, qubit 2 experiences spontaneous decay, and the QRS undergoes various lossy channels

Notice that our system preserves the following symmetry. If we change $\hat{\sigma}_x^i \rightarrow -\hat{\sigma}_x^i$ and $(\hat{a} + \hat{a}^\dagger) \rightarrow -(\hat{a} + \hat{a}^\dagger)$, the Hamiltonian in Eq. (6.2) remains unchanged, i.e., the system is symmetric with respect to an inversion of the pseudo-spin operator, and the field quadrature. The reader is advised to refer to Chap. 3 about the underlying $\mathbb{Z}_2$ symmetry of the quantum Rabi model. We remark that there exists a parity symmetry $\hat{\Pi} = -e^{i\pi\hat{a}^\dagger\hat{a}}\hat{\sigma}_z^p\hat{\sigma}_z^1\hat{\sigma}_z^2$ such that $[\hat{H}, \hat{\Pi}] = 0$. As the result, $\hat{H}$ and $\hat{\Pi}$ can be simultaneously diagonalized by $|\phi_j\rangle$, where $\hat{\Pi}|\phi_j\rangle = \pm|\phi_j\rangle$, and $\hat{H}|\phi_j\rangle = \epsilon_j|\phi_j\rangle$, $\forall j$. Also, the parity symmetry establishes selection rules in our system. For instance, states with different parities can only be connected via an interaction that breaks the symmetry, while states with the same parity can be connected by an interaction that preserves the symmetry.

6.2 Equilibrium and Nonequilibrium Properties

Here, we present the main features of our model shown in Fig. 6.1. First, we show the equilibrium properties and demonstrate the performance of a tunable strong qubit–qubit interaction in the context of superconducting circuits. Second, we show how our system might be used as a primitive unit cell for cavity mediated excitation transfer in the USC regime and its connection with cavity-enhanced exciton transport with organic matter and optical microcavities [8–10].

6.2.1 Effective Two-Body Interaction in Dispersive Limit

In this subsection, we show both analytically and numerically that the catalytic QRS can be treated as a mediating quantum bus between two qubits, providing an effective two-body interaction. Analogous to the effective interaction mediated by a resonator bus in cavity/circuit QED configurations [14–16], we demonstrate an effective qubit–qubit interaction as a second-order process, due to a dispersive coupling between the qubits and QRS.

To attain the effective qubit–qubit interaction, we consider a dispersive approach beyond the rotating wave approximation (RWA). Unlike the dispersive theory of Ref. [17], we consider the QRS as a whole, and all relevant timescales are compared with all available energy levels, $\omega_{jk} = \omega_j - \omega_k$, where ω_j satisfy $\hat{H}_p|j\rangle = \omega_j|j\rangle$. The strategy of the entire derivation, leading to the effective Hamiltonian, is as follows. We start from considering a system, comprising of N qubits that interact with a QRS, i.e.,

$$\hat{H}' = \hat{H}_0 + \hat{H}_I, \qquad (6.3)$$

where $\hat{H}_0 = \hat{H}_p + \sum_{n=1}^{N} \frac{\omega_{qn}}{2}\hat{\sigma}_z^n$, and the interaction Hamiltonian $\hat{H}_I = \sum_{n=1}^{N} g_n\hat{\sigma}_x^n(\hat{a} + \hat{a}^\dagger)$. We note that the expression $\hat{H}'$ is the same as Eq. (6.2), with the presence of N qubits inside the resonator. At the end, we would restrict ourselves with $N = 2$ qubits case, and restricting to the ground and first excited states of the central QRS, thereby arriving at the effective interacting Hamiltonian for the system shown in Fig. 6.1.

Projecting the total system onto the central QRS eigenbasis, we have $\hat{H}_p = \sum_{j=0}^{\infty} \omega_j |j\rangle\langle j|$. Using the completeness relation and projecting the interaction term onto the QRS eigenbases, we then obtain

$$\hat{H}_I = \sum_{n=1}^{N} g_n \hat{\sigma}_x^n \sum_{j,k=0}^{\infty} \left[|k\rangle\langle k|(\hat{a} + \hat{a}^\dagger)|j\rangle\langle j|\right] = \sum_{n,j,k>j} g_n \hat{\sigma}_x^n \left[\chi_{kj}|k\rangle\langle j| + \text{h. c.}\right], \tag{6.4}$$

where $\chi_{kj} = \langle k|(\hat{a} + \hat{a}^\dagger)|j\rangle$. Without applying the RWA, we obtain the interaction Hamiltonian in the interaction picture as

$$\tilde{\hat{H}}_I(t) = \sum_{n,j,k>j} g_n \left[\chi_{jk}e^{i\Delta_{kj}^n t}\hat{\sigma}_+^n |j\rangle\langle k| + \chi_{kj}e^{i\delta_{kj}^n t}\hat{\sigma}_+^n |k\rangle\langle j| + \right.$$
$$\left. \chi_{jk}e^{-i\delta_{kj}^n t}\hat{\sigma}_-^n |j\rangle\langle k| + \chi_{kj}e^{-i\Delta_{kj}^n t}\hat{\sigma}_-^n |k\rangle\langle j|\right]. \tag{6.5}$$

Here, $\Delta_{kj}^n = \omega_{qn} - \omega_{kj}$, $\delta_{kj}^n = \omega_{qn} + \omega_{kj}$, and $\tilde{\hat{H}}$ refers to Hamiltonian in the interaction picture. The relevant timescales in our system dynamics come from various energy level differences of the QRS and two qubits frequencies. Here, we are interested in the dispersive limit where the N qubits frequencies are far off-resonant with the lowest QRS transition frequency (the difference between the ground and first excited states of the QRS). In this case, fast oscillatory dynamics can be averaged out to zero and thus only slow evolving dynamics contribute to the overall system dynamics. Hence, we are interested in the time average of a measurable physical operator $\hat{O}(t)$ [18, 19] as

$$\overline{\hat{O}}(t) \equiv \int_{-\infty}^{\infty} f(t - t')\hat{O}(t')dt', \tag{6.6}$$

where we suppose that $f(t)$ is a real function and it has a unit area. As the result, the usual time-ordered evolution operator, satisfying the Schrödinger equation,

$$i\frac{\partial}{\partial t}\hat{U}(t, t_0) = \tilde{\hat{H}}_I(t)\hat{U}(t, t_0), \tag{6.7}$$

can now be rewritten as

$$i\frac{\partial}{\partial t}\overline{\hat{U}(t, t_0)} = \hat{\mathcal{H}}_{\text{eff}}(t)\overline{\hat{U}(t, t_0)}, \tag{6.8}$$

when we involve the time-averaging operator, defined in Eq. (6.6). From Eqs. (6.7)–(6.8), a general expression for $\hat{\mathcal{H}}_{\text{eff}}(t)$ can be written as

$$\hat{\mathcal{H}}_{\text{eff}}(t) = \overline{[\tilde{\hat{H}}_I(t)\hat{U}(t,t_0)]}\,\overline{[\hat{U}(t,t_0)]}^{-1},\tag{6.9}$$

where

$$\hat{U}(t,t_0) = \hat{\mathcal{T}}\exp\left[-i\int_{t_0}^{t}\tilde{\hat{H}}_I(t')dt'\right].\tag{6.10}$$

Here, $\hat{\mathcal{T}}$ represents a time-ordering operator. Let us remark the following important notion of unitarity in the operator $\hat{U}(t,t_0)$. Though it is unitary, its time averaged operator is in general not. Thus, $\hat{\mathcal{H}}_{\text{eff}}(t)$ shown above is not Hermitian, since we have traced out the high-frequency part/s. However, the effective Hamiltonian for the unitary part of the evolution is uniquely given by its Hermitian part [18]:

$$\hat{H}_{\text{eff}}(t) = \frac{1}{2}\{\hat{\mathcal{H}}_{\text{eff}}(t) + \hat{\mathcal{H}}_{\text{eff}}(t)^{\dagger}\}.\tag{6.11}$$

Up to the first-order Taylor series expansion, $\hat{U}(t,t_0) \approx \hat{1} + \hat{U}_1(t)$, where $\hat{U}_1(t) = -i\int_{t_0}^{t}dt'\tilde{\hat{H}}_I(t')$, Eq. (6.11) can be rewritten as

$$\hat{H}_{\text{eff}}(t) = \overline{\tilde{\hat{H}}_I(t)} + \frac{1}{2}\left(\overline{[\tilde{\hat{H}}_I(t),\hat{U}_1(t)]} - \overline{[\tilde{\hat{H}}_I(t),\hat{U}_1(t)]}\right).\tag{6.12}$$

Explicitly, $\hat{U}_1(t)$ has the following expression:

$$\hat{U}_1(t) = -i\sum_{n,j,k>j}g_n\left[\frac{\chi_{jk}}{i\Delta_{kj}^n}\left(e^{i\Delta_{kj}^n t}-1\right)\hat{\sigma}_+^n|j\rangle\langle k| + \frac{\chi_{kj}}{i\delta_{kj}^n}\left(e^{i\delta_{kj}^n t}-1\right)\hat{\sigma}_+^n|k\rangle\langle j|\right.$$
$$\left. + \frac{\chi_{jk}}{-i\delta_{kj}^n}\left(e^{-i\delta_{kj}^n t}-1\right)\hat{\sigma}_-^n|j\rangle\langle k| + \frac{\chi_{kj}}{-i\Delta_{kj}^n}\left(e^{-i\Delta_{kj}^n t}-1\right)\hat{\sigma}_-^n|k\rangle\langle j|\right].\tag{6.13}$$

By substituting Eqs. (6.5) and (6.13) into Eq. (6.12) and transforming back to the Schrödinger picture, we arrive at the following effective Hamiltonian:

$$\hat{H}_{\text{eff}} = \hat{H}_0 + \frac{1}{2}\sum_{n,n'}g_ng_{n'}\sum_{j,k>j}|\chi_{jk}|^2 \times$$

$$\left[\left\{\left(\frac{1}{\Delta_{kj}^n}-\frac{1}{\delta_{kj}^{n'}}\right)\hat{\sigma}_+^n\hat{\sigma}_+^{n'} + \left(\frac{1}{\Delta_{kj}^n}+\frac{1}{\Delta_{kj}^{n'}}\right)\hat{\sigma}_+^n\hat{\sigma}_-^{n'}\right.\right.$$

$$\left.\left. - \left(\frac{1}{\delta_{kj}^n}+\frac{1}{\delta_{kj}^{n'}}\right)\hat{\sigma}_-^n\hat{\sigma}_+^{n'} + \left(\frac{1}{\Delta_{kj}^{n'}}-\frac{1}{\delta_{kj}^n}\right)\hat{\sigma}_-^n\hat{\sigma}_-^{n'}\right\}|j\rangle\langle j|\right.$$

$$\left. + \left\{\left(\frac{1}{\delta_{kj}^n}-\frac{1}{\Delta_{kj}^{n'}}\right)\hat{\sigma}_+^n\hat{\sigma}_+^{n'} + \left(\frac{1}{\delta_{kj}^{n'}}+\frac{1}{\delta_{kj}^n}\right)\hat{\sigma}_+^n\hat{\sigma}_-^{n'}\right.\right.$$

$$-\left(\frac{1}{\Delta_{kj}^n} + \frac{1}{\Delta_{kj}^{n'}}\right)\hat{\sigma}_-^n\hat{\sigma}_+^{n'} + \left(\frac{1}{\delta_{kj}^{n'}} - \frac{1}{\Delta_{kj}^n}\right)\hat{\sigma}_-^n\hat{\sigma}_-^{n'}\Bigg\}|k\rangle\langle k|\Bigg]. \quad (6.14)$$

In deriving the above expression, we have neglected oscillating terms that are proportional to $\exp(\pm i\omega_{kk'}t)$. This is supplemented by the assumption that

$$\omega_{kk'} \gg \{g_ng_{n'}|\chi_{jk}|^2/\Delta_{kj}^n, g_ng_{n'}|\chi_{jk}|^2/\delta_{kj}^n\}, \quad (6.15)$$

which can be guaranteed with proper choice of the QRS parameters. When we take $N = 2$ case, we attain

$$\hat{H}_{\text{eff}} = \hat{H}_0 + \sum_{j,k>j}|\chi_{jk}|^2 \times \Bigg[\Bigg\{\sum_{n=1}^{2}\left(\frac{g_n^2}{\Delta_{kj}^2}\hat{\sigma}_+^n\hat{\sigma}_-^n - \frac{g_n^2}{\delta_{kj}^n}\hat{\sigma}_-^n\hat{\sigma}_+^n\right)$$

$$+ \frac{g_1g_2}{2}\left(\frac{1}{\Delta_{kj}^1} + \frac{1}{\Delta_{kj}^2} - \frac{1}{\delta_{kj}^1} - \frac{1}{\delta_{kj}^2}\right)\hat{\sigma}_x^1\hat{\sigma}_x^2\Bigg\}|j\rangle\langle j|$$

$$+ \Bigg\{\sum_{n=1}^{2}\left(\frac{g_n^2}{\delta_{kj}^2}\hat{\sigma}_+^n\hat{\sigma}_-^n - \frac{g_n^2}{\Delta_{kj}^n}\hat{\sigma}_-^n\hat{\sigma}_+^n\right)$$

$$+ \frac{g_1g_2}{2}\left(\frac{1}{\delta_{kj}^1} + \frac{1}{\delta_{kj}^2} - \frac{1}{\Delta_{kj}^1} - \frac{1}{\Delta_{kj}^2}\right)\hat{\sigma}_x^1\hat{\sigma}_x^2\Bigg\}|k\rangle\langle k|\Bigg]. \quad (6.16)$$

When we confine ourselves to the two lowest energy levels of the QRS (i.e., $j = 0, k = 1$), we arrive at

$$\hat{H}_{\text{eff}} = \hat{H}_0 + \frac{1}{2}|\chi_{01}|^2[\hat{S}_{12}|1\rangle\langle 1| - \hat{S}_{12}|0\rangle\langle 0|] = \hat{H}_0 + \frac{1}{2}|\chi_{01}|^2\hat{S}_{12}\otimes\hat{Z}_p, \quad (6.17)$$

where $|\chi_{01}|^2 = |\langle 0|(\hat{a} + \hat{a}^\dagger)|1\rangle|^2$, $\hat{Z}_p = |1\rangle\langle 1| - |0\rangle\langle 0|$, $\hat{S}_{12} = g_1g_2(1/\delta_{10}^1 + 1/\delta_{10}^2 - 1/\Delta_{10}^1 - 1/\Delta_{10}^2)\hat{\sigma}_x^1 \otimes \hat{\sigma}_x^2 + \sum_{n=1}^{2}2g_n^2(\hat{\sigma}_+^n\hat{\sigma}_-^n/\Delta_{10}^n - \hat{\sigma}_-^n\hat{\sigma}_+^n/\delta_{10}^n)$. We would like to remark that the Schrieffer–Wolff transformation [20], $e^{\hat{S}}\hat{H}e^{-\hat{S}}$, has been applied to the total Hamiltonian, Eq. (6.2), with the following non-Hermitian operator (see Appendix C for the detailed derivation of the effective Hamiltonian via the Schrieffer–Wolff transformation, and how we obtain the expression of $\hat{S}$)

$$\hat{S} = \sum_{n=1}^{2}\sum_{j,k>j}g_n\left[\frac{\chi_{jk}}{\Delta_{kj}^n}\hat{\sigma}_+^n|j\rangle\langle k| - \frac{\chi_{kj}}{\Delta_{kj}^n}\hat{\sigma}_-^n|k\rangle\langle j| + \frac{\chi_{kj}}{\delta_{kj}^n}\hat{\sigma}_+^n|k\rangle\langle j| - \frac{\chi_{jk}}{\delta_{kj}^n}\hat{\sigma}_-^n|j\rangle\langle k|\right], \quad (6.18)$$

and the time-averaging operator method produce the same effective qubit–qubit Hamiltonian, up to a constant term.

First, we have seen that the entire system Hamiltonian equation (6.2) can be approximated by the effective Hamiltonian equation (6.17), when the QRS interacts

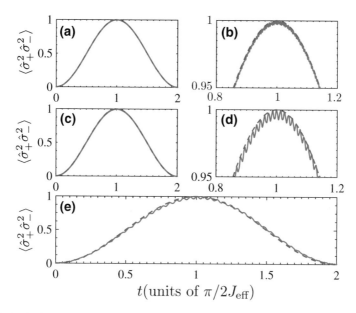

Fig. 6.2 Excitation number of the second qubit as a function of time. **a** $g_p/\omega_{cav} = 0.1$, $\omega_p = 0.8\omega_{cav}$, $\omega_{q1} = \omega_{q2} = 0.2\omega_{cav}$, and $g_1 = g_2 = 0.02\omega_{cav}$. These parameters lead to an effective qubit–qubit coupling strength $2J_{eff} = 0.00176\omega_{cav}$. **b** Enlarged portion of (**a**). **c** $g_p/\omega_{cav} = 0.3$, $\omega_p = 0.8\omega_{cav}$, $\omega_{q1} = \omega_{q2} = 0.2\omega_{cav}$, and $g_1 = g_2 = 0.02\omega_{cav}$. These parameters lead to an effective qubit–qubit coupling strength $2J_{eff} = 0.00267\omega_{cav}$. **d** Enlarged portion of (**c**). **e** $g_p/\omega_{cav} = 0.5$, $\omega_p = 0.8\omega_{cav}$, $\omega_{q1} = \omega_{q2} = 0.2\omega_{cav}$, and $g_1 = g_2 = 0.02\omega_{cav}$. These parameters lead to an effective qubit–qubit coupling strength $2J_{eff} = 0.00573\omega_{cav}$. In all the figures, blue (continuous) lines are evolution outcome under the full Hamiltonian, equation (6.2), and red (dashed) lines are evolution outcome under the effective Hamiltonian, equation (6.17)

dispersively with the two qubits. In Fig. 6.2, we plot the excitation number of the second qubit $\langle \hat{\sigma}_+^2 \hat{\sigma}_-^2 \rangle$ as a function of time in accordance with both of the Hamiltonians, given that initial system state is at $|0_p\rangle|\uparrow_1\downarrow_2\rangle^1$. We use red (dashed) lines for the effective Hamiltonian evolution and blue (continuous) lines for the evolution under the exact full Hamiltonian. We see that the two results match pretty well in small g_p/ω_{cav} parameter regime (cf. Fig. 6.2a–d). However, when g_p/ω_{cav} is reasonably large as in Fig. 6.2e, a clear deviation from the full Hamiltonian dynamics is then observed. The reason is that at this coupling strength the QRS gap is closer to the qubits energy splitting such that the QRS can now be excited. Moreover, the assumption we make in arriving at the effective Hamiltonian: $\omega_{kk'} \gg \{g_n g_{n'}|\chi_{jk}|^2/\Delta_{kj}^n, g_n g_{n'}|\chi_{jk}|^2/\delta_{kj}^n\}$ is not true any more. The second observation is that fast oscillations in the full Hamiltonian evolution become apparent with increase in g_p/ω_{cav}, while they are smeared out in the effective Hamiltonian evolutions, because we have employed the time-

---

[1] p subscript represents the QRS, while subscript 1 and 2 means qubit 1 and 2. We will adopt this representation without specifying the subscripts.

averaging operators of the form in Eq. (6.6). From here onwards, we use $|\downarrow\rangle = |g\rangle$ and $|\uparrow\rangle = |e\rangle$, interchangeably.

We next consider the simplest scenario, where two identical qubits interact with a QRS. In this case, the Hilbert space of the whole system is spanned by tensor products of the QRS eigenbasis $\{|j_p\rangle\}$ ($j = 0, 1, ..., \infty$), and the symmetric Dicke states $\{|D_{N,k}\rangle\}$ with the number of qubits, $N = 2$ and excitation number $k = 0, 1, 2, \cdots$. Namely, $|D_{2,0}\rangle = |gg\rangle$, $|D_{2,1}\rangle = (|eg\rangle + |ge\rangle)/\sqrt{2}$, and $|D_{2,2}\rangle = |ee\rangle$. Figure 6.3a shows the lowest energy states as a function of the qubits energy gap $\omega_{q1} = \omega_{q2} = \Delta$. Notice that varying the qubits energy gap does not change the underlying $\mathbb{Z}_2$ symmetry. As the energy gap approaches the QRS energy, precisely at $\Delta = 0.6042\omega_{cav}$, the energy spectrum shows avoided level crossing between the states with the same parity. In addition, the spectrum shows a straight line representing the state $|0_p\rangle|\psi_-\rangle$, where $|\psi_-\rangle = (|eg\rangle - |ge\rangle)/\sqrt{2}$ is the singlet state that does not couple with the QRS. However, this state does not appear in spectroscopic measurements [21, 22].

The lowest energy states shown in Fig. 6.3a are linear superposition of the QRS and two qubits states $\{|j_p\rangle|D_{N,k}\rangle\}$, which are the approximated states of the total Hamiltonian equation (6.2), when the latter is truncated to the basis defined by the states $\{|0_p\rangle, |1_p\rangle\} \otimes \{|gg\rangle, |ge\rangle, |eg\rangle, |ee\rangle\}$. At the first avoided level crossing with the resonance condition $\Delta = 0.6042\omega_{cav}$, both the qubits and QRS are maximally entangled. They can be approximated by $|G_\pm\rangle \approx |0_p\rangle|D_{2,1}\rangle \pm |1_p\rangle|D_{2,0}\rangle)/\sqrt{2}$. In particular, states $|0\rangle|D_{2,1}\rangle$ and $|1\rangle|D_{2,0}\rangle$ exhibit population inversion as shown in Fig. 6.3b. Notice that the QRS energy spectrum (see Fig. 6.3c) is recovered, while the dashed vertical line depicts the $\Delta = 0$ in Fig. 6.3a.

When the two qubits are not identical, which is always the case for superconducting qubits, we obtain different results in Fig. 6.4a as compared to Fig. 6.3a. We do observe an avoided level crossing at $\Delta/\omega_{cav} = 0.2$, without the presence of the non-interacting state $|0_p\rangle|\psi_-\rangle$, represented by a straight line in Fig. 6.3a. Figure 6.4a shows the energy spectrum of the total system, comprising of the QRS and two qubits, while qubit 1 energy gap is fixed at $\omega_{q1} = 0.2\omega_{cav}$ and qubit 2 energy gap is varied from $\omega_{q2} \in [0, 1.2]\omega_{cav}$. The enlarged diagram is also shown in Fig. 6.4b, where we identify the two states approximately as $|0_p\rangle|\psi_+\rangle$ and $|0_p\rangle|\psi_-\rangle$, where $|\psi_\pm\rangle = (|eg\rangle \pm |ge\rangle)/\sqrt{2}$. At the avoided level crossing, these two states have an energy gap $\Delta_{gap} = 2J_{eff}$. To show this numerically, we plot the populations $P_{|0_p\rangle|eg\rangle}$ and $P_{|0_p\rangle|ge\rangle}$ as a function of time with a prior initial state $|\bar{\psi}\rangle = |0_p\rangle|eg\rangle$. The result can be seen in Fig. 6.4c. The von Neumann entropy, $S(\rho_A)$, between the QRS and two qubits is also plotted with yellow solid line in the same figure, where we see negligible entanglement between the two subsystems. Hence, it guarantees that the QRS is not excited during the qubit–qubit interaction. Furthermore, we observe fast oscillations in the two populations in Fig. 6.4c and d. One can show that these fast oscillations have a negligible contribution for the weak qubit–QRS coupling strength, i.e., $g_1, g_2 \ll g_p$, and the dynamics predicted from the full Hamiltonian, Eq. (6.2), and the effective Hamiltonian, Eq. (6.17), match perfectly well.

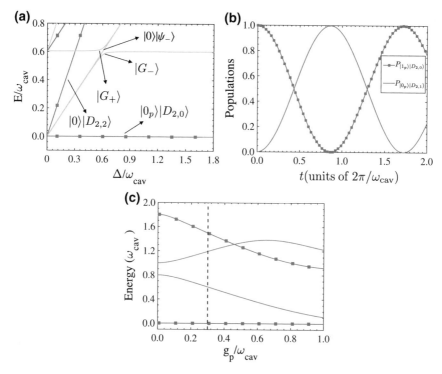

Fig. 6.3 **a** Energy spectrum ($\hbar = 1$) from the Hamiltonian (6.2) for identical qubits, $\omega_{q1} = \omega_{q2} = \Delta$. Red (squared) lines stand for states with even parity ($p = +1$) and yellow (continuous) lines stand for states with odd parity ($p = -1$), where p is the eigenvalue of the parity operator P. **b** Population inversion between states $|0_p\rangle|D_{2,1}\rangle$ and $|1_p\rangle|D_{2,0}\rangle$ at the resonance condition $\Delta = 0.6042\omega_{\text{cav}}$. The numerical calculations for (**a**) and (**b**) are done with parameters $\omega_p = 0.8\omega_{\text{cav}}$, $g_p = 0.3\omega_{\text{cav}}$, and $g_1 = g_2 = 0.02\omega_{\text{cav}}$. **c** Energy spectrum for a single QRS (see Eq. (6.1)), with parameters $\omega_p = 0.8\omega_{\text{cav}}$, as a function of the light–matter coupling. Blue (squared) lines stand for states with even parity and green (continuous) lines for states with odd parity associated with the parity operator $\hat{\Pi}_p = -e^{i\pi\hat{a}^\dagger\hat{a}}\hat{\sigma}_z^p$. The vertical line stands for $g_p/\omega_{\text{cav}} = 0.3$, corresponding to $\Delta = 0$ in (**a**). The states indicated here are not the actual eigenstates of the total Hamiltonian, equation (6.2). They are approximate states that are calculated in the truncated subspace $\{|0_p\rangle, |1_p\rangle\} \otimes \{|gg\rangle, |ge\rangle, |eg\rangle, |ee\rangle\}$

6.2.2 Entangling Two Qubits

In a realistic setup, observing Rabi oscillations between the two states $|eg\rangle$ and $|ge\rangle$ can be achieved by making use of Gaussian and Stark control pulses as described in Ref. [16], where a strong qubit–qubit interaction is mediated via a cavity bus. The protocol for entangling the qubits is as follows. (1) Let the system to cool down to its ground state $|0_p\rangle|gg\rangle$. (2) We apply a Gaussian π pulse acting upon the qubit 1 in order to prepare the state $|0_p\rangle|eg\rangle$. (3) A Stark pulse is applied to the qubit 2, bringing the qubits into resonance for a time Δt. In this way, the strong qubit–qubit interaction will induce the desired population transfer between states $|eg\rangle$ and $|ge\rangle$.

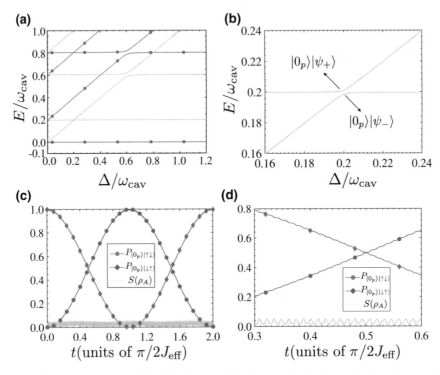

Fig. 6.4 a Energy spectrum ($\hbar = 1$) of the Hamiltonian equation (6.2) as a function of qubit 2 energy, $\omega_{q2} = \Delta$. Here, we consider parameters $\omega_p = 0.8\omega_{cav}$, $g_p = 0.3\omega_{cav}$, $\omega_{q1} = 0.2\omega_{cav}$, and $g_1 = g_2 = 0.02\omega_{cav}$. Red (squared) lines stand for states with even parity and yellow (continuous) lines stand for states with odd parity. **b** First avoided level crossing at $\Delta = 0.2\omega_{cav}$ is enlarged. **c** Population inversion between states $|0_p\rangle|eg\rangle$, blue (circle) line, and $|0_p\rangle|ge\rangle$, red (diamond) line, at the first avoided level crossing is shown. The yellow (continuous) line shows the von Neumann entropy between the QRS and two qubits, indicating negligible entanglement between the two subsystems. **d** We enlarge fast oscillations seen in (**c**). The states indicated in (**a**) and (**b**) are not the actual eigenstates of the total Hamiltonian, equation (6.2). They are approximate states in dispersive limit

   In our proposal, the time for entangling two qubits is drastically reduced with increasing g_p/ω_{cav}. For instance, if we consider the cavity frequency $\omega_{cav} = 2\pi \times 8$ GHz for the USC system as in Ref. [23], our model predicts effective qubit–qubit coupling strengths of $2J_{eff} = 2\pi \times (21, 28, 46)$ MHz for $g_p/\omega_{cav} = (0.3, 0.4, 0.5)$, respectively. These values lead to times of $t = \pi/(4J_{eff}) \approx (11, 9, 5)$ ns, which scale similar to or better than the time needed to perform a controlled-phase gate with fast and resonant gates in new-generation circuit QED setups [24–26]. Indeed, one could increase the effective qubit–qubit coupling strength by increasing the ratio g_p/ω_{cav} in the QRS. For instance, with coupling strengths of $g_p/\omega_{cav} = (0.6, 0.8)$, and cavity frequency of $\omega_{cav} = 2\pi \times 8$ GHz, one can reach effective coupling strengths of about $2J_{eff} = 2\pi \times (77, 160)$ MHz. These values lead to times of $t = \pi/(4J_{eff}) \approx (3, 1.5)$ ns. However, the latter coupling parameters g_p/ω_{cav} violate

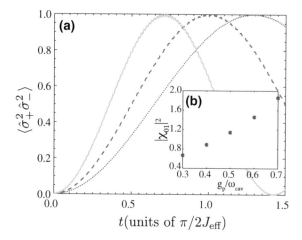

Fig. 6.5 **a** Excitation number of the qubit 2 as a function of time. Yellow (continuous) line stands for $g_p = 0.4\omega_{cav}$, red (dashed) line stands for $g_p = 0.3\omega_{cav}$, and blue (dotted) line stands for $g_p = 0.2\omega_{cav}$. In all these cases, we consider $2J_{eff} = 0.00267\omega_{cav}$ corresponding to parameters $\omega_p = 0.8\omega_{cav}$, $g_p = 0.3\omega_{cav}$, $\omega_{q1} = \omega_{q2} = 0.2\omega_{cav}$, and $g_1 = g_2 = 0.02\omega_{cav}$. **b** (inset) The matrix element $|\chi_{01}|^2$ is plotted against g_p/ω_{cav}

the dispersive interaction between the QRS and the two qubits, since the QRS gap closes as g_p/ω_{cav} increases. A single restriction of our proposal is to work within a parameter range for g_p/ω_{cav} that enables the dispersive interaction. We stress that our scheme does not require a tunable qubit–cavity coupling as in the case of circuit QED-based ultrafast gates [4, 7].

Improvement in excitation transfer is achieved with increase in g_p/ω_{cav} as shown in Fig. 6.5, where we study the excitation number of the qubit 2 as a function of g_p/ω_{cav}, with the initial input state $|0\rangle|eg\rangle$. The larger g_p/ω_{cav} is, the faster the rate of excitation transfer becomes. There are two physical reasonings for this. The first one is that an increase in the matrix element $|\chi_{01}|^2$ with increasing g_p/ω_{cav}, as seen in Fig. 6.5b. The second one is that the hybridized cavity frequency is reduced, thus reducing the energy cost of the virtual process mediated by the QRS. We note that similar improvement of excitation transfer in a linear resonator bus can, in principle, be attainable by changing the resonator frequency or its impedance [14–16]. The only difference with our proposal is that we are using a nonlinear QRS bus here.

6.2.3 Excitation Transfer Between Qubits

Motivated by recent developments in cavity-enhanced energy and charge transport with organic matter in optical microcavities [8–11] in the strong coupling regime, in addition to the results obtained in Fig. 6.5, we next study the nonequilibrium case, where the qubit 1 is continuously pumped with incoherent radiation, and observe the excitation transfer towards the qubit 2. The natural question from this simple nonequilibrium setup is to ask how the total system behaves in the long run. We investigate steady-state excitation transfer towards a spontaneously emitting acceptor (modeled with qubit 2), when a donor (qubit 1) is incoherently pumped at rate γ_{pump}. Both the donor and acceptor are mediated by a lossy QRS. Our idea follows from the photovoltaic models of Refs. [27, 28], where energy is assumed to dissipate in a reaction center at the final stage of the energy transfer chain. The reaction center physically corresponds to an acceptor molecule that undergoes a charge separation event, thus producing electric work. The extent of charge separation in a reaction center is proportional to the population of acceptor excited state, which is the observable we monitor in our numerical calculations. We stress that our study represents a minimal model for a photovoltaic cell that needs to be further explored in order to propose a realistic photovoltaic array.

Since our system governed by Eq. (6.2) has a large anharmonicity, whose eigenvalues and eigenvectors are defined by $H|\phi_j\rangle = \epsilon_j|\phi_j\rangle$, we need to consider the coloured nature of baths and the hybridization of the qubit–cavity operators [29–33]. With these considerations, we follow the microscopic master equation already discussed in the Chap. 5 and Appendix B, which reads

$$\dot{\rho}(t) = -i[\hat{H}, \hat{\rho}(t)] + \sum_{n=x_1,x_2,x_p,z_p,b} \mathcal{L}_n \hat{\rho}(t). \tag{6.19}$$

Here, $\mathcal{L}_{x_1}$ and $\mathcal{L}_{x_2}$ are Liouvillian superoperators describing the incoherent pumping upon qubit 1 (γ_{pump}) and the spontaneous emission of qubit 2 (γ_{out}). Moreover, we include loss mechanisms acting on the QRS via transversal noise (γ_x), longitudinal noise (γ_z), and noise acting on the field quadrature (γ_{cav}), through Liouvillian superoperators $\mathcal{L}_{x_p}$, $\mathcal{L}_{z_p}$, and $\mathcal{L}_b$. In particular, $\mathcal{L}_{x_1}\hat{\rho}(t) = \sum_{j,k>j} \Gamma_{x_1}^{jk} \mathcal{D}[|\phi_k\rangle\langle\phi_j|]\hat{\rho}(t)$ and $\mathcal{L}_\sigma\hat{\rho}(t) = \sum_{j,k>j} \Gamma_\sigma^{jk} \mathcal{D}[|\phi_j\rangle\langle\phi_k|]\hat{\rho}(t)$ for $\sigma = x_2, x_p, z_p, b$, where $\mathcal{D}[\mathcal{O}]\hat{\rho}(t) = \frac{1}{2}[2\hat{O}\hat{\rho}(t)\hat{O}^\dagger - \hat{\rho}(t)\hat{O}^\dagger\hat{O} - \hat{O}^\dagger\hat{O}\hat{\rho}(t)]$. Here, the frequency dependent rates are $\Gamma_{x_1}^{jk} = \gamma_{\text{pump}}\frac{\epsilon_{kj}}{\omega_{q1}}|\langle\phi_j|\hat{\sigma}_x^1|\phi_k\rangle|^2$, $\Gamma_{x_2}^{jk} = \gamma_{\text{out}}\frac{\epsilon_{kj}}{\omega_{q2}}|\langle\phi_j|\hat{\sigma}_x^2|\phi_k\rangle|^2$, $\Gamma_{x_p}^{jk} = \gamma_x\frac{\epsilon_{kj}}{\omega_p}|\langle\phi_j|\hat{\sigma}_x^P|\phi_k\rangle|^2$, $\Gamma_{z_p}^{jk} = \gamma_z\frac{\epsilon_{kj}}{\omega_p}|\langle\phi_j|\hat{\sigma}_z^P|\phi_k\rangle|^2$, and $\Gamma_b^{jk} = \gamma_{\text{cav}}\frac{\epsilon_{kj}}{\omega_{\text{cav}}}|\langle\phi_j|(\hat{a}+\hat{a}^\dagger)|\phi_k\rangle|^2$, where $\epsilon_{kj} = \epsilon_k - \epsilon_j$.

We employ the superspace operator method [34] to find the steady-state solution of the system density matrix. If the system of interest belongs to a Hilbert space of dimension $\dim(\mathcal{H}) = d$, the master equation in the superspace method reads

$$|\dot{\rho}\rangle = \left[-i\,\hat{H}\otimes\hat{\mathbb{I}} + i\,\hat{\mathbb{I}}\otimes\hat{H}^T + \sum_{n=x_1,x_2,x_p,z_p,b} \tilde{\mathcal{L}}_n\right]|\rho\rangle, \tag{6.20}$$

where each superoperator $\tilde{\mathcal{L}}_n \propto \frac{1}{2}[2\hat{O}\otimes\hat{O}^* - \hat{\mathbb{I}}\otimes\hat{O}^T\hat{O}^* - \hat{O}^\dagger\hat{O}\otimes\hat{\mathbb{I}}]$. Here, superscript ^{T} means transpose operation while * means complex conjugate operation, and † stands for conjugate transpose. Notice that the superspace dimension is $\dim(\mathcal{S}) = d^2$, and $|\dot{\rho}\rangle = d|\rho\rangle/dt$. In the master equation above, $\hat{\mathbb{I}}$ is the superspace identity and we

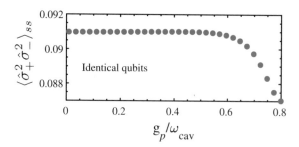

Fig. 6.6 Steady-state excitation number of qubit 2 versus g_p/ω_{cav}. In this simulation, $\omega_{q1} = \omega_{q2} = 0.2\omega_{cav}$, $\omega_p = 0.8\omega_{cav}$, $g_1 = g_2 = 10^{-2}\omega_{cav}$ are considered. The pumping and loss rates are $\gamma_{pump} = \gamma_x = \gamma_z = \gamma_{cav} = 10^{-2}\omega_{cav}$, and $\gamma_{out} = 10^{-1}\omega_{cav}$. We note that a Fock space of $N = 8$ is enough to assure convergence for each value of the ratio g_p/ω_{cav}

take into account the transpose and conjugate of system operators. The steady-state solution $(\hat{\rho}_{ss})$ is found numerically under the condition $|\dot{\rho}\rangle = 0$, which, in another words, implies finding the eigenstate of $\mathcal{L} = -i\,\hat{H} \otimes \hat{\mathbb{I}} + i\,\hat{\mathbb{I}} \otimes \hat{H}^T + \sum_n \tilde{\mathcal{L}}_n$, with zero eigenvalue, i.e., $\mathcal{L}\,|\rho\rangle_{ss} = 0$.

We study the steady-state excitation transferred to qubit 2 $(\langle \sigma_+^2 \sigma_-^2 \rangle_{ss})$ for two cases, namely, identical qubit case and nonidentical one. For the identical qubit case, the steady-state population of the qubit 2, obtained from the master equation, Eq. (6.19) with the ab initio Hamiltonian (6.2), exhibits a flat behavior until $g/\omega_{cav} \approx 0.6$ as shown in Fig. 6.6. For the chosen parameters mentioned in Fig. 6.6, the value $g/\omega_{cav} \approx 0.6$ establishes a limit where both qubits depart from the dispersive regime with the QRS. Beyond this point, physics is not captured by virtual excitation of the QRS, since the latter and the qubits become resonance.

The flat behavior when decreasing the g/ω_{cav} has an intuitive explanation if we study the system eigenstates of the effective qubit–qubit Hamiltonian (6.17). There, the effective qubit–qubit coupling strength J_{eff} is very small compared with the qubits frequencies ω_{q1} and ω_{q2} such that one can perform the rotating wave approximation in the qubit–qubit interaction

$$\hat{H}_{eff} \approx \sum_{n=1}^{2} \frac{\tilde{\omega}_{qn}}{2}\hat{\sigma}_z^n + J_{eff}(\hat{\sigma}_+^1 \hat{\sigma}_-^2 + \hat{\sigma}_-^1 \hat{\sigma}_+^2), \qquad (6.21)$$

where $\tilde{\omega}_{qn} = \omega_{qn} + |\chi_{01}|^2 g_n^2 (1/\Delta_{10}^n + 1/\delta_{10}^n)$ and $J_{eff} = \frac{1}{2}|\chi_{01}|^2 g_1 g_2 (1/\delta_{10}^1 + 1/\delta_{10}^2 - 1/\Delta_{10}^1 - 1/\Delta_{10}^2)$. The eigenvalues and eigenstates of Hamiltonian (6.21) are $E_G = -\frac{1}{2}\sqrt{\tilde{\omega}_{q1} + \tilde{\omega}_{q2}}$, $E_1 = -\frac{1}{2}\sqrt{4J_{eff}^2 + (\tilde{\omega}_{q1} - \tilde{\omega}_{q2})^2}$, $E_2 = \frac{1}{2}\sqrt{4J_{eff}^2 + (\tilde{\omega}_{q1} - \tilde{\omega}_{q2})^2}$, $E_3 = \frac{1}{2}\sqrt{\tilde{\omega}_{q1} + \tilde{\omega}_{q2}}$ and $|G\rangle = |gg\rangle$, $|E_1\rangle = -\sin(\theta/2)|eg\rangle + \cos(\theta/2)|ge\rangle$, $|E_2\rangle = \cos(\theta/2)|eg\rangle + \sin(\theta/2)|ge\rangle$, $|E_3\rangle = |ee\rangle$, with $\tan(\theta) = 2J_{eff}/(\tilde{\omega}_{q1} - \tilde{\omega}_{q2})$. It is apparent that for identical qubit frequencies $\omega_{q1} = \omega_{q2}$ and qubit–QRS coupling strengths $g_1 = g_2$, the eigenstates of the joint qubit–qubit system do not depend on the effective coupling strength J_{eff}. This implies that neither the matrix elements of

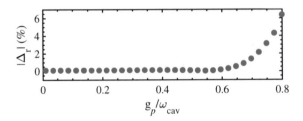

Fig. 6.7 The absolute value of the relative difference Δ_r in percentage, computed numerically from the ab initio Hamiltonian (6.2) ($\langle \hat{\sigma}_+^2 \hat{\sigma}_-^2 \rangle_{ss}^{ab\ initio}$) and from the effective two-qubit Hamiltonian (6.21) ($\langle \hat{\sigma}_+^2 \hat{\sigma}_-^2 \rangle_{ss}^{effective}$), is plotted against g_p/ω_{cav}, with the same parameters of Fig. 6.6

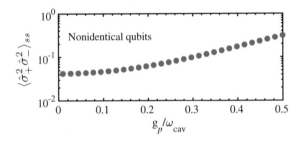

Fig. 6.8 Steady-state excitation number of qubit 2 versus g_p/ω_{cav}, for the case of nonidentical qubits. In this simulation, we used parameters $\omega_{q1} = 0.2\omega_{cav}$, $\omega_{q2} = 0.19\omega_{cav}$, $\omega_p = 0.8\omega_{cav}$, $g_1 = g_2 = 10^{-2}\omega_{cav}$, $\gamma_{out} = 10^{-4}\omega_{cav}$, $\gamma_{pump} = \gamma_x = \gamma_z = \gamma_{cav} = 10^{-2}\omega_{cav}$

operators appearing in the microscopic master equation for the effective two-qubit system

$$\dot{\hat{\rho}}_Q(t) = -i[\hat{H}_{eff}, \hat{\rho}_Q(t)] + \sum_{n=x_1,x_2} \mathcal{L}_n \hat{\rho}_Q(t), \tag{6.22}$$

where $\hat{\rho}_Q$ describes the two qubits density matrix, nor the energy differences will have an influence on J_{eff}, since the dominating frequency scale is ω_{q1} (ω_{q2}).

Notice that the Liouvillian superoperator $\mathcal{L}_n$ needs to be evaluated in terms of the effective qubit–qubit basis described above. We perform numerical calculations to obtain the steady-state solution $\langle \hat{\sigma}_+^2 \hat{\sigma}_-^2 \rangle_{ss}$ computed numerically from the master equation (6.19) with the ab initio Hamiltonian (6.2) ($\langle \hat{\sigma}_+^2 \hat{\sigma}_-^2 \rangle_{ss}^{ab\ initio}$) and the one from the master equation (6.22) with the effective two-qubit Hamiltonian (6.21) ($\langle \hat{\sigma}_+^2 \hat{\sigma}_-^2 \rangle_{ss}^{effective}$). Figure 6.7 shows the absolute value of the relative difference $\Delta_r = 1 - \langle \hat{\sigma}_+^2 \hat{\sigma}_-^2 \rangle_{ss}^{effective}/\langle \hat{\sigma}_+^2 \hat{\sigma}_-^2 \rangle_{ss}^{ab\ initio}$ in percentage, for the same parameters used in Fig. 6.6.

For nonidentical qubit case where $\omega_{q1} \neq \omega_{q2}$, the result of the steady-state mean value of the excitation number of qubit 2, obtained from the master equation (6.19) with the ab initio Hamiltonian (6.2), is shown in Fig. 6.8. As the coupling strength of the QRS enters the USC regime, $0.1 \lesssim g_p/\omega_{cav} < 1$, we see a striking one order of magnitude increase in the excitation transfer as g_p/ω_{cav} increases from

10^{-2} (the strong coupling regime) to 0.5 (the USC regime). It is noteworthy that the nonidentical qubits case represents a more realistic approach towards donor–acceptor organic photovoltaic complex, where donor and acceptor molecules are not identical.

6.3 Transmon-Based Implementation

Here, in this section, we try to propose a realistic circuit QED setup by making use of a flux qubit galvanically embedded inside a $\lambda/2$ microwave resonator for implementing the QRS. Two additional transmon circuits can be positioned at the resonator edges, where the resonator voltage is maximum. In this way, possible crosstalk between on-chip flux lines can be avoided. We also note that since transmons are many-level systems rather than qubits, so we need to carry out additional analysis in order to demonstrate our proposal. The Hamiltonian for a single transmon device reads

$$\hat{H}_T = 4E_C(\hat{N} - N_g)^2 - E_J \cos(\hat{\theta}), \tag{6.23}$$

where E_C is the charging energy, E_J is the Josephson energy, and N_g is the effective offset charge of the device in units of $2e$. Also, $\hat{N}$ represents the number of Cooper pairs transferred to the superconducting island, and $\hat{\theta}$ stands for the gauge-invariant phase difference across the Josephson junction. Figure 6.9 shows the lowest three energy levels of the transmon as a function of the ratio E_J/E_C, and $N_g = 0$. Here, we take realistic transmon charging energy $E_C/\hbar = 2\pi \times 0.31$ GHz [25]. Also, if we consider a cavity frequency $\omega_{\text{cav}} = 2\pi \times 16$ GHz, which can be achieved with current circuit QED setups [35], we obtain $E_C/\hbar = 0.0194\omega_{\text{cav}}$. In this simulation, we have fixed the ground state energy to the zero. Notice that for parameters of the QRS $\omega_p = \omega_{\text{cav}}$ and $g_p = 0.3\omega_{\text{cav}}$, one can demonstrate that the frequency difference between the ground and excited state of the QRS is about $\omega_{10} = 1.4\omega_{\text{cav}}$. Also, at $E_J/E_C \approx 49$ ($E_J/\hbar = 2\pi \times 15.3$ GHz [25]), the absolute anharmonicity of the transmon is $\alpha = E_{10} - E_{21} = 0.0223\omega_{\text{cav}} = 2\pi \times 356.8$ MHz.

Based on the above results, we have performed simulations that include two nonidentical transmons coupled to the QRS, and we truncate each transmon Eq. (6.23) to its three lowest energy levels. The Hamiltonian that describes this situation reads

$$\hat{H} = \sum_{\ell=1}^{2} \sum_{j=0}^{2} E_j^{(\ell)} |j_\ell\rangle\langle j_\ell| + \hat{H}_p + \sum_{\ell=1}^{2} g_\ell \hat{N}_\ell(\hat{a} + \hat{a}^\dagger), \tag{6.24}$$

where $E_j^{(\ell)}$ and $|j_\ell\rangle$ stand for the jth frequency and eigenstate associated with the ℓth transmon, respectively. $\hat{H}_p$ is the Hamiltonian of the QRS, g_ℓ and $\hat{N}_\ell$ are the ℓth transmon–QRS coupling strength and the Copper pairs number, respectively. Fig. 6.10 shows the results as a function of the ratio E_J/E_C for transmon 2, and for fixed ratio E_J/E_C for the transmon 1 taken from the simulation performed in

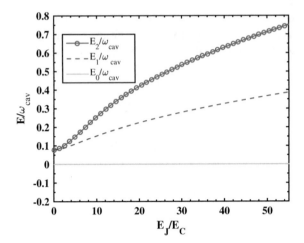

Fig. 6.9 Energy spectrum vs E_J/E_C for the transmon device described by the Hamiltonian (6.23). Here, the charging energy $E_C/\hbar = 0.0194\omega_{cav}$

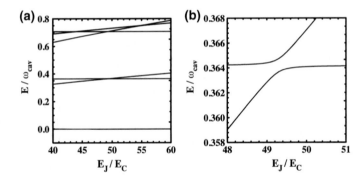

Fig. 6.10 **a** Transmon–QRS–transmon energy spectrum for the Hamiltonian (6.24) versus the ratio E_J/E_C for the transmon 2. **b** Enlarged first avoided level crossing in (**a**). Here, the parameters for transmon 1 are fixed to $E_C = 0.011\omega_{cav}$, $E_J/E_C \approx 49$, and the parameters for the QRS are $g_p/\omega_{cav} = 0.3$ and $\omega_p = \omega_{cav}$

Fig. 6.9. We see that at the resonance condition $E_J/E_C \approx 49$, the lowest avoided level crossing appears at $E/\omega_{cav} \approx 0.3646$ (see Fig. 6.10b), which is below from the QRS excited state frequency, $\omega_{10} = 1.4\omega_{cav}$. Thus, the QRS remains at its ground state. This resembles the avoided crossing that appears in Fig. 6.4b. Also, it shows evidence of the effective transmon–transmon interaction mediated by QRS, such that the transmon-based implementation would allow us to simulate our parity-preserving light–matter system. This is analogous to the transmon-based implementation of effective qubit–qubit interaction in circuit QED setups [16]. Notice that Fig. 6.10a also shows additional level crossings at higher energies that are caused by extra allowed transitions in the effective transmon–transmon interaction.

6.4 Summary and Discussions

In this chapter, we have investigated the equilibrium and nonequilibrium dynamics of two qubits interacting via a QRS. We have demonstrated a possibility of using the QRS as a quantum bus to mediate a strong and tunable qubit–qubit interaction. In particular, two-qubit entanglement time can be reduced with the increase qubit–cavity coupling within the QRS. We also highlight that the manipulation of the qubits energy relaxes the requirement of complex flux qubit architectures, and thereby without requiring tunable qubit–cavity coupling to perform qubit–qubit interaction as in Ref. [4]. Hence, our theoretical proposal could be implemented in a circuit QED setup with existing technologies. In particular, one can consider a flux qubit coupled galvanically to an on-chip $\lambda/2$ transmission line resonator to form the QRS [36]. Two additional transmon qubits are positioned at the resonator edges, where the resonator voltage is maximum. In this way, possible magnetic crosstalk between on-chip flux lines can be avoided. Furthermore, since transmons are many-level systems rather than qubits, additional analysis is required. Nevertheless, our proposal can also be extended to multilevel systems as shown in Sect. 6.3. We like to emphasize that our system might have applications beyond quantum information processing. In particular, we have seen an improvement of excitation transfer between two nonidentical two-level systems when g_p/ω_{cav} is increased, while one of them is incoherently pumped, thus providing the possibility of a minimal model for a photovoltaic cell. This work has led to the publication [37]: T. H. Kyaw, S. Allende, L.-C. Kwek and G. Romero, *Parity-preserving light–matter system mediates effective two-body interactions*, *Quantum Sci. Technol.* **2**, 025007 (2017). http://iopscience.iop.org/article/10. 1088/2058-9565/aa701c/meta.

References

1. Wang YD, Zhang P, Zhou DL, Sun CP (2004) Fast entanglement of two charge-phase qubits through nonadiabatic coupling to a large Josephson junction. Phys Rev B 70:224515
2. Wang Y-D, Kemp A, Semba K (2009) Coupling superconducting flux qubits at optimal point via dynamic decoupling with the quantum bus. Phys Rev B 79:024502
3. Chen C-Y (2011) Geometric phase gate based on both displacement operator and squeezed operators with a superconducting circuit quantum electrodynamics. Commun Theor Phys 56(1):91
4. Romero G, Ballester D, Wang YM, Scarani V, Solano E (2012) Ultrafast quantum gates in circuit QED. Phys Rev Lett 108:120501
5. Kyaw TH, Herrera-Martí DA, Solano E, Romero G, Kwek L-C (2015a) Creation of quantum error correcting codes in the ultrastrong coupling regime. Phys Rev B 91:064503
6. Blais A, Huang R-S, Wallraff A, Girvin SM, Schoelkopf RJ (2004) Cavity quantum electrodynamics for superconducting electrical circuits: an architecture for quantum computation. Phys Rev A 69:062320
7. Peropadre B, Forn-Díaz P, Solano E, García-Ripoll JJ (2010) Switchable ultrastrong coupling in circuit QED. Phys Rev Lett 105:023601

8. Orgiu E, George J, Hutchison JA, Devaux E, Dayen JF, Doudin B, Stellacci F, Genet C, Schachenmayer J, Genes C, Pupillo G, Samori P, Ebbesen TW (2015) Conductivity in organic semiconductors hybridized with the vacuum field. Nat Mater 14(11):1123
9. Schachenmayer J, Genes C, Tignone E, Pupillo G (2015) Cavity-enhanced transport of excitons. Phys Rev Lett 114:196403
10. Feist J, Garcia-Vidal FJ (2015) Extraordinary exciton conductance induced by strong coupling. Phys Rev Lett 114:196402
11. Mazzeo M, Genco A, Gambino S, Ballarini D, Mangione F, Di Stefano O, Patanè S, Savasta S, Sanvitto D, Gigli G (2014) Ultrastrong light-matter coupling in electrically doped microcavity organic light emitting diodes. Appl Phys Lett 104(23)
12. Rabi II (1936) On the process of space quantization. Phys Rev 49:324
13. Braak D (2011) Integrability of the Rabi model. Phys Rev Lett 107:100401
14. Zheng S-B, Guo G-C (2000) Efficient scheme for two-atom entanglement and quantum information processing in cavity QED. Phys Rev Lett 85:2392
15. Blais A, Gambetta JM, Wallraff A, Schuster DI, Girvin SM, Devoret MH, Schoelkopf RJ (2007) Quantum-information processing with circuit quantum electrodynamics. Phys Rev A 75:032329
16. Majer J, Chow JM, Gambetta JM, Koch J, Johnson BR, Schreier JA, Frunzio L, Schuster DI, Houck AA, Wallraff A, Blais A, Devoret MH, Girvin SM, Schoelkopf RJ (2007) Coupling superconducting qubits via a cavity bus. Nature 449(7161):443
17. Zueco D, Reuther GM, Kohler S, Hänggi P (2009) Qubit-oscillator dynamics in the dispersive regime: analytical theory beyond the rotating-wave approximation. Phys Rev A 80:033846
18. James DF, Jerke J (2007) Effective hamiltonian theory and its applications in quantum information. Can J Phys 85(6):625
19. Gamel O, James DFV (2010) Time-averaged quantum dynamics and the validity of the effective Hamiltonian model. Phys Rev A 82:052106
20. Schrieffer JR, Wolff PA (1966) Relation between the Anderson and Kondo Hamiltonians. Phys Rev 149:491
21. Bretheau L, Girit CO, Pothier H, Esteve D, Urbina C (2013) Exciting Andreev pairs in a superconducting atomic contact. Nature 499(7458):312
22. Janvier C, Tosi L, Bretheau L, Girit ÇÖ, Stern M, Bertet P, Joyez P, Vion D, Esteve D, Goffman MF, Pothier H, Urbina C (2015) Coherent manipulation of Andreev states in superconducting atomic contacts. Science 349(6253):1199
23. Forn-Díaz P, Lisenfeld J, Marcos D, García-Ripoll JJ, Solano E, Harmans CJPM, Mooij JE (2010) Observation of the Bloch-Siegert shift in a qubit-oscillator system in the ultrastrong coupling regime. Phys Rev Lett 105:237001
24. Haack G, Helmer F, Mariantoni M, Marquardt F, Solano E (2010) Resonant quantum gates in circuit quantum electrodynamics. Phys Rev B 82:024514
25. Saira O-P, Groen JP, Cramer J, Meretska M, de Lange G, DiCarlo L (2014) Entanglement genesis by ancilla-based parity measurement in 2D circuit QED. Phys Rev Lett 112:070502
26. Chen Y, Neill C, Roushan P, Leung N, Fang M, Barends R, Kelly J, Campbell B, Chen Z, Chiaro B et al (2014) Qubit architecture with high coherence and fast tunable coupling. Phys Rev Lett 113:220502
27. Zhang Y, Wirthwein A, Alharbi FH, Engel GS, Kais S (2016) Dark states enhance the photocell power via phononic dissipation. Phys Chem Chem Phys 18:31845
28. Dorfman KE, Voronine DV, Mukamel S, Scully MO (2013) Photosynthetic reaction center as a quantum heat engine. Proc Natl Acad Sci 110(8):2746
29. De Liberato S, Gerace D, Carusotto I, Ciuti C (2009) Extracavity quantum vacuum radiation from a single qubit. Phys Rev A 80:053810
30. Beaudoin F, Gambetta JM, Blais A (2011) Dissipation and ultrastrong coupling in circuit QED. Phys Rev A 84:043832
31. Ridolfo A, Leib M, Savasta S, Hartmann MJ (2012) Photon blockade in the ultrastrong coupling regime. Phys Rev Lett 109:193602

32. Sete EA, Gambetta JM, Korotkov AN (2014) Purcell effect with microwave drive: Suppression of qubit relaxation rate. Phys Rev B 89:104516
33. Govia LCG, Wilhelm FK (2015) Unitary-feedback-improved qubit initialization in the dispersive regime. Phys Rev Appl 4:054001
34. Navarrete-Benlloch C (2015) Open systems dynamics: simulating master equations in the computer. arXiv:1504.05266
35. Forn-Díaz P, At Institute for Quantum Computing University of Waterloo (private communication)
36. Niemczyk T, Deppe F, Huebl H, Menzel EP, Hocke F, Schwarz MJ, Garcia-Ripoll JJ, Zueco D, Hummer T, Solano E, Marx A, Gross R (2010) Circuit quantum electrodynamics in the ultrastrong-coupling regime. Nat Phys 6:772
37. Kyaw TH, Allende S, Kwek L-C, Romero G (2017) Parity-preserving light-matter system mediates effective two-body interactions. Quantum Sci Technol 2(2):025007
38. Douçot B, Feigel'man MV, Ioffe LB, Ioselevich AS (2005) Protected qubits and chern-simons theories in Josephson junction arrays. Phys Rev B 71:024505

Chapter 7
Conclusion and Future Work

I have discovered a truly remarkable proof of this theorem which this margin is too small to contain.

— Pierre de Fermat

Unlike current classical computers, a quantum computer promises efficient processing capability for certain computational problems [1–3]. In order to realize efficient quantum circuitry that outperforms its classical counterpart, it is therefore important to profit from the unique quantum mechanical features, offered by the Nature, that optimize and enhance computations. At present, we are living in an exciting time, experiencing such a paradigm in many experimental expenditures all around the world. There are numerous platforms toward practical quantum computing such as superconducting circuits, ion traps, nitrogen-vacancy centers, photonic waveguides, just to name a few.

7.1 Conclusion

In the present thesis, we have presented various quantum computing schemes within the framework of superconducting circuits architecture, while we try to profit from the ultrastrong light–matter interaction. The USC interaction is not commonly available in Mother Nature. One way to attain the interaction is via the use of circuit QED. Therefore, in the beginning, we have outlined the general overview of the circuit QED community in the context of both strong coupling (Jaynes–Cummings model) and USC (quantum Rabi model) regimes. We gave a general introduction to various

© Springer Nature Switzerland AG 2019
T. H. Kyaw, *Towards a Scalable Quantum Computing Platform in the Ultrastrong Coupling Regime*, Springer Theses, https://doi.org/10.1007/978-3-030-19658-5_7

types of artificial qubits available in the literature. Among many different types, we focus on the flux and transmon qubits throughout the thesis.

The main theme of the thesis is to provide quantum computing framework in the USC limit and possibly extend the framework to many-qubit setup in the superconducting circuit architecture. First, we start with the ultrafast two-qubit gate in the USC regime based on our published work [4], with which we were able to generate smallest quantum error correcting code that is capable to both detect and correct an arbitrary quantum error. We then move on to generate larger quantum error correcting code widely known as the Steane code. We demonstrate that we can create both QECCs in nanosecond timescale which is far faster compared to present decoherence lifetime of the artificial qubits in the lab. With the help of the Monte Carlo simulations, we give some reliable thresholds within which one can create the two QECCs with some acceptable quantum errors.

In Chap. 5, we demonstrate the possibility to use a flux qubit in resonator at the USC regime as a quantum memory cell. This is based on our work in Ref. [5, 6]. We achieve this by observing that there is no energy level crossing present for the ground and first excited states, in the quantum Rabi model energy spectrum from the strong coupling all the way to the ultra-strong coupling regime. We also explain that for the certain type of superconducting qubit, the ground and first excited states in the USC regime are robust against certain quantum noise, and thus the encoding in the USC can attain three order of magnitude improvement over the usual strong coupling qubit. With the use of tunable qubit-resonator coupling in the superconducting circuit, we propose to adiabatically tune the light–matter coupling from the strong coupling to the USC, whenever there is a necessity to store important quantum information. During the retrieval process, one has to adiabatically tune the coupling back to the strong coupling limit. We can do this with a very good fidelity provided the adiabatic theorem holds during the storage and retrieval processes. We have also shown that we can store not only pure quantum states but also arbitrary superposed states as well as two-qubit entangled states in our quantum memory cells. We then use the open quantum system approach to analyze the same two processes and we find that we can get the fidelity of 0.9939 for certain types of common experimental errors in circuit QED. Lastly, we tried to extend our memory cell into a two-dimensional square lattice configuration, with the help of SQUID in every node of the lattice.

Chapter 6 mainly deals with the means to overcome some of the experimental challenges imposed by earlier proposals in the USC circuits, where the tunable qubit-resonator coupling is used extensively (Chaps. 4 and 5). In this chapter, we provide an alternative way to avoid using the tunable interaction, yet achieve the similar performance of the two-qubit entangling gate operation as the USC one [7]. We achieve this by inserting a USC system in the middle of the two strongly coupled qubits, which can be seen in our work of Ref. [8]. The central USC polariton would then mediate as an effective qubit–qubit interaction for the two qubits. We find that the qubit–qubit interaction gets stronger with the increase of the light–matter coupling in the central USC system, as long as the dispersive limit criteria is satisfied. We also study the steady-state excitation transport from the left qubit to the right one within the framework of the open quantum system. The promising results seem to indicate the development of efficient solar cells in the near future.

In conclusion, we have demonstrated simple quantum computing tasks in the USC limit where the two-qubit entangling gates and quantum memory devices were proposed. We hope that with the latest experimental developments in circuit QED platform, our theoretical proposals in this thesis would soon be useful in the near future.

7.2 Future Work

Looking forward, we dream to achieve large-scale USC quantum computing, by joining together our quantum memory devices in a regular lattice configuration as shown in Fig. 7.1. As the first step, we have proposed the constituent of the large-scale quantum memory shown in Fig. 7.1a in the Chap. 5. Each node is constituted with our memory cell element. In addition, each node in the network is connected to a SQUID that allows to selectively switch on/off interaction between neighboring microwave cavities [7, 10], in order to implement quantum state transfer processes within the same layer.

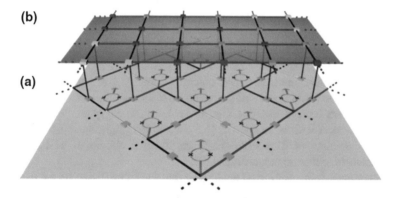

Fig. 7.1 A scalable quantum memory [5]. **a** The light–matter interface operating at the ultra-strong coupling regime may be envisioned as a set of microwave cavities connected, at the nodes, by SQUID devices that allow to switch on/off the cavity–cavity interaction. Notice that each cavity is represented with different colors (red, black, blue, and green) that stand for different lengths to assure the manipulation of specific pairwise interactions (see Supplementary information). In addition, on each edge of the memory array, there is a memory cell made of a USC entity (blue square) to store an arbitrary quantum information in a specific location. **b** Integrated quantum processor. A 2D cavity grid with a qubit distribution (rectangular boxes) represented in various colors is shown here. It was previously shown in Ref. [9] that such a cavity grid may provide a scalable fault-tolerant quantum computing architecture with minimal swapping overhead. Data transfer between the two layers may be done via cavity buses (vertical black colored lines connecting layer a and b)

The scale-up problem is still a big issue for both experimentalists and theorists. To address the issue in details, we need to make sure that when a single photon propagates along the 2D platform, its quality does not deteriorate with the distance it travels, and it only interacts with the specific memory cell within the 2D layout. This grand problem can then be divided into the following smaller ones:

1. Microwave photon generation has to be studied properly such that the generated photon can then be absorbed [11] by the memory cell with near unity fidelity.
2. Dynamics of single photon propagation [12] has to be analyzed in details for the configuration shown in Fig. 7.1 or that of the configurations associated with the fanout quantum random access memory (QRAM) and the Bucket-brigade QRAM [13, 14].
3. Near-term practical quantum computing circuits are to be developed to reliably control and scale-up the entire architecture by minimizing environmental noise and crosstalk. For that, we are hoping that experimentalists will make some breakthroughs.

To elaborate the last point (3), there is one recent exciting development in miniaturizing microwave circulator onto the superconducting chip. The circulator is an invaluable device for noise management and signal routing for circuit QED experiments. Currently, commercial ferrite junction circulators are bulky whose physical dimension is close to the microwave wavelength applied. To shrink the size down to micrometer scale, there are recent theoretical proposals using the quantum Hall effect [15] and active devices [16]. Their corresponding experimental realizations can be seen in Ref. [17] and Ref. [18], respectively.

Ultimately, we would like to achieve a multilayer circuit architecture, where a quantum processor layer [9] interfaces with the proposed memory layer as shown in Fig. 7.1. Remarkably, the recent IBM quantum computing research patent [19] on the modular superconducting circuit architecture has some similarity to our proposal, Fig. 7.1 [5]. We believe that this kind of architecture can pave the way for the implementation of a scalable QRAM [13, 14], which might benefit from the fast storage and readout performances of superconducting circuits as well as various quantum machine learning algorithms [20].

Another system that we intend to study in the near future based on the thesis is a realistic solar cell model involving many qubits in the USC regime, similar to the one shown in Fig. 6.1. There are many approaches to study this complicated model. One way is to numerically simulate the N-body USC system in brute force, by adopting the current techniques we used to study the simple model of Fig. 6.1. One alternative means is to incorporate nonequilibrium many-body theory of quantum systems [21], in particular, by making use of the Keldysh Green function techniques. In this way, one should be able to take into consideration the inter-qubit interactions as well as the interaction between the entire system and an environment in proper. The hope is that one would be able to go beyond the usual Born-Markovian approximation [22] for the N-body USC system. On the way to this ultimate goal, one simplest scenario is to study a single USC system couples to non-Markovian bath, which we might be able to approach it from the Floquet stroboscopic divisibility [23] that we have proposed recently. Hence, some interesting precursors [24, 25] of the non-Markovian dynamics can still be studied in a small USC system.

Hence, we believe that an exciting journey is ahead of us and this thesis is the first step in this exciting direction.

References

1. Papageorgiou A, Traub JF (2013) Measures of quantum computing speedup. Phys Rev A 88(2):022316
2. Rønnow TF, Wang Z, Job J, Boixo S, Isakov SV, Wecker D, Martinis JM, Lidar DA, Troyer M (2014) Defining and detecting quantum speedup. Science 345(6195):420
3. Heim B, Rønnow TF, Isakov SV, Troyer M (2015) Quantum versus classical annealing of ising spin glasses. Science 348(6231):215
4. Kyaw TH, Herrera-Martí DA, Solano E, Romero G, Kwek L-C (2015a) Creation of quantum error correcting codes in the ultrastrong coupling regime. Phys Rev B 91:064503
5. Kyaw TH, Felicetti S, Romero G, Solano E, Kwek L-C (2015b) Scalable quantum memory in the ultrastrong coupling regime. Sci Rep 5:8621
6. Kyaw TH, Felicetti S, Romero G, Solano E, Kwek L-C (2014) $\mathbb{Z}_2$ quantum memory implemented on circuit quantum electrodynamics. Proc SPIE 9225:92250B
7. Romero G, Ballester D, Wang YM, Scarani V, Solano E (2012) Ultrafast quantum gates in circuit QED. Phys Rev Lett 108:120501
8. Kyaw TH, Allende S, Kwek L-C, Romero G (2017) Parity-preserving light-matter system mediates effective two-body interactions. Quantum Sci Technol 2(2):025007
9. Helmer F, Mariantoni M, Fowler AG, von Delft J, Solano E, Marquardt F (2009) Cavity grid for scalable quantum computation with superconducting circuits. Eur Phys Lett 85(5):50007
10. Felicetti S, Sanz M, Lamata L, Romero G, Johansson G, Delsing P, Solano E (2014) Dynamical Casimir effect entangles artificial atoms. Phys Rev Lett 113(9):093602
11. Wang Y, Minář J, Hétet G, Scarani V (2012) Quantum memory with a single two-level atom in a half cavity. Phys Rev A 85(1):013823
12. Longo P, Schmitteckert P, Busch K (2009) Dynamics of photon transport through quantum impurities in dispersion-engineered one-dimensional systems. J Opt A: Pure Appl Opt 11(11):114009
13. Giovannetti V, Lloyd S, Maccone L (2008a) Quantum random access memory. Phys. Rev. Lett 100:160501
14. Giovannetti V, Lloyd S, Maccone L (2008b) Architectures for a quantum random access memory. Phys Rev A 78:052310
15. Viola G, DiVincenzo DP (2014) Hall effect gyrators and circulators. Phys Rev X 4(2):021019
16. Kerckhoff J, Lalumière K, Chapman BJ, Blais A, Lehnert KW (2015) On-chip superconducting microwave circulator from synthetic rotation. Phys Rev Appl 4(3):034002
17. Mahoney AC, Colless JI, Pauka SJ, Hornibrook JM, Watson JD, Gardner GC, Manfra MJ, Doherty AC, Reilly DJ (2017) On-chip microwave quantum hall circulator. Phys Rev X 7(1):011007
18. Chapman BJ, Rosenthal EI, Kerckhoff J, Moores BA, Vale LR, Mates JAB, Hilton GC, Lalumiere K, Blais A, Lehnert KW (2017) Widely tunable on-chip microwave circulator for superconducting quantum circuits. Phys Rev X 7(4):041043
19. Chow JM, Gambetta JM, Rothwell MB, Rozen JR (2018). Modular array of vertically integrated superconducting qubit devices for scalable quantum computing. US Patent App 15/871,443
20. Childs AM (2009) Quantum algorithms: equation solving by simulation. Nat Phys 5(12):861

21. Stefanucci G, Van Leeuwen R (2013) Nonequilibrium many-body theory of quantum systems: a modern introduction. Cambridge University Press
22. Chakraborty A, Sensarma R (2018) Power-law tails and non-Markovian dynamics in open quantum systems: an exact solution from Keldysh field theory. Phys Rev B 97(10):104306
23. Bastidas VM, Kyaw TH, Tangpanitanon J, Romero G, Kwek L-C, Angelakis DG (2018) Floquet stroboscopic divisibility in non-Markovian dynamics. New J Phys 20:093004
24. Tan J, Kyaw TH, Yeo Y (2010) Non-Markovian environments and entanglement preservation. Phys Rev A 81(6):062119
25. Kyaw TH, Bastidas VM, Tangpanitanon J, Romero G, Kwek L-C (2018) Dynamical quantum phase transitions and non-markovian dynamics. arXiv:1811.04621
26. Douçot B, Feigel'man MV, Ioffe LB, Ioselevich AS (2005) Protected qubits and chern-simons theories in Josephson junction arrays. Phys Rev B 71:024505

Appendix A
Derivation of the USC Evolution Operator

It has been shown that magnetic fluxes $\hat{\Phi}_{ex_1}^j$ can tune the coefficients c_x^j and c_z^j (see Fig. 4.2) to attain the longitudinal coupling with $c_x^j \approx 0$ and $c_z^j \approx 1$ [1, 2], which is an ideal setting for the pairwise cluster state generation in an ultrafast timescale. For each mode ℓ, we define a displacement operator

$$\hat{\mathcal{D}}_\ell(\sum_j \kappa_j \hat{\sigma}_z^j) = \exp[(\sum_j \kappa_j \hat{\sigma}_z^j)\hat{a}_\ell^\dagger - (\sum_j \kappa_j \hat{\sigma}_z^j)\hat{a}_\ell], \qquad (A.1)$$

with $\kappa_j = g_j/\omega$ and $\omega_\ell = \omega$ since we consider a collective resonator mode at a degeneracy point [3]. In addition, for all the modes within the manifold $\mathcal{M}$, we define a collective displacement operator $\hat{\mathcal{D}}(\xi) = \prod_{\ell \in \mathcal{M}} e^{\hat{\xi}\hat{a}_\ell^\dagger - \hat{\xi}^\dagger \hat{a}_\ell}$, where $\hat{\xi} = \left(\sum_j \kappa_j \hat{\sigma}_z^j\right)$. By transforming the original Hamiltonian, Eq. (4.16) with the above operator, we obtain

$$\hat{H}' = \hat{\mathcal{D}}^\dagger(\hat{\xi})\hat{\mathcal{D}}(\hat{\xi})\hat{H}\hat{\mathcal{D}}^\dagger(\hat{\xi})\hat{\mathcal{D}}(\hat{\xi}) = \hat{\mathcal{D}}^\dagger(\hat{\xi})[\omega \sum_\ell \hat{a}_\ell^\dagger \hat{a}_\ell - \omega M \hat{\xi}^2]\hat{\mathcal{D}}(\hat{\xi}), \qquad (A.2)$$

where M is the dimension of $\mathcal{M}$. The evolution operator associated with the new $\hat{H}'$ is given by

$$\hat{U}(t) = \hat{U}_0(t)e^{i\omega t M \hat{\xi}^2}e^{-i\omega t \hat{\mathcal{D}}^\dagger(\hat{\xi})(\sum_\ell \hat{a}_\ell^\dagger \hat{a}_\ell)\hat{\mathcal{D}}(\hat{\xi})} \qquad (A.3)$$

$$= \hat{U}_0(t)e^{i\hat{\xi}^2 M(\omega t - \sin(\omega t))}\prod_\ell e^{-i\omega t \hat{a}_\ell^\dagger \hat{a}_\ell}\hat{\mathcal{D}}_\ell[\hat{\xi}(t)],$$

with $\hat{U}_0(t) = \exp[-it \sum_j \frac{\omega_q^j}{2}\hat{\sigma}_z^j]$ and $\hat{\mathcal{D}}_\ell[\hat{\xi}(t)] = \hat{\mathcal{D}}_\ell((1 - e^{i\omega t})\hat{\xi})$. After an evolution time $t = 2\pi n/\omega$, we have

$$\hat{U}(2\pi n/\omega) = \hat{U}_0(2\pi n/\omega)e^{i\hat{\xi}^2 M(2\pi n)}\prod_\ell e^{-2\pi n i \hat{a}_\ell^\dagger \hat{a}_\ell}, \qquad (A.4)$$

© Springer Nature Switzerland AG 2019
T. H. Kyaw, *Towards a Scalable Quantum Computing Platform in the Ultrastrong Coupling Regime*, Springer Theses, https://doi.org/10.1007/978-3-030-19658-5

where n is an integer multiple. Since our protocol constitutes of pairwise qubits,

$$\hat{U}(2\pi/\omega) \approx \exp[\frac{-i\pi}{\omega}(\omega_q^i \hat{\sigma}_z^i + \omega_q^j \hat{\sigma}_z^j) \times$$
$$\exp[i4\pi n M((\kappa_i^2 + \kappa_j^2)\frac{I}{2} + \kappa_i \kappa_j \hat{\sigma}_z^i \hat{\sigma}_z^j)]. \qquad (A.5)$$

Thus, we have

$$\hat{U}_{CZ} = \hat{\mathcal{U}} \times \exp\left[\frac{-i\pi}{4}(\hat{\sigma}_z^i + \hat{\sigma}_z^j)\right] \times$$
$$\exp\left[4\pi i M\left((\kappa_i^2 + \kappa_j^2)\frac{I}{2} + \kappa_i \kappa_j \hat{\sigma}_z^i \hat{\sigma}_z^j\right)\right], \qquad (A.6)$$

where $\hat{\mathcal{U}} = \exp\left[\frac{-i\pi}{4}\left[\left(\frac{4\omega_q^i - \omega}{\omega}\right)\hat{\sigma}_z^i + \left(\frac{4\omega_q^j - \omega}{\omega}\right)\hat{\sigma}_z^j\right]\right]$ and $n = 1$. To perform the controlled-phase gate operation with a maximum fidelity, we require that both $\kappa_i^2 + \kappa_j^2 = \frac{1}{8nM}$ and $\kappa_i \kappa_j = \frac{1}{16nM}$ are satisfied. In other words, we need $\kappa_i + \kappa_j = \frac{1}{2\sqrt{nM}}$.

Appendix B
Microscopic Derivation of an Open Quantum System in the USC Regime

B.1 Master Equation via a Microscopic Derivation

In this section, we adopt a microscopic derivation formalism from the theory of open quantum systems [4] to arrive at a proper quantum master equation for the ultrastrong coupled system. Let us consider a scenario where a system, consisting of a qubit ultrastrongly coupled to a resonator, weakly interacts with a bath environment. In the interaction picture, the system evolution can be written as

$$\frac{d\hat{\rho}_S(t)}{dt} = -\int_0^t d\tau \ \mathrm{Tr}_B[\hat{H}_I(t), [\hat{H}_I(\tau), \hat{\rho}(\tau)]], \tag{B.1}$$

where $\hat{\rho}_S$ represents the system density operator, the subscript B represents the bath, and $\hat{\rho}$ is the combined system and bath density matrix. Furthermore, we involve the Born approximation where the system–bath interaction is weak. That means the bath degrees of freedom are negligibly affected by the weak interaction. Therefore, the total system at time t can be approximated by $\hat{\rho}(t) \approx \hat{\rho}_S(t) \otimes \hat{\rho}_B$. Finally, we arrive at

$$\frac{d\hat{\rho}_S(t)}{dt} = -\int_0^t d\tau \ \mathrm{Tr}_B[\hat{H}_I(t), [\hat{H}_I(\tau), \hat{\rho}_S(\tau) \otimes \hat{\rho}_B]]. \tag{B.2}$$

Furthermore, we assume environmental excitations decay over time and cannot be resolved. This assumption is know as the Markovian approximation, with which we arrive at the Redfield master equation

$$\frac{d\hat{\rho}_S(t)}{dt} = -\int_0^t d\tau \ \mathrm{Tr}_B[\hat{H}_I(t), [\hat{H}_I(\tau), \hat{\rho}_S(t) \otimes \hat{\rho}_B]]. \tag{B.3}$$

We then substitute τ by $t - \tau$ and change the upper limit of the integral to ∞. This is allowable provided the integrand vanishes sufficiently fast for $\tau \gg \tau_B$, the

© Springer Nature Switzerland AG 2019
T. H. Kyaw, *Towards a Scalable Quantum Computing Platform in the Ultrastrong Coupling Regime*, Springer Theses,
https://doi.org/10.1007/978-3-030-19658-5

timescale over which the state of the system varies is large compared to the timescale over which the bath correlation functions decay. Thus, we arrive at the Markovian quantum master equation

$$\frac{d\hat{\rho}_S(t)}{dt} = -\int_0^\infty d\tau \ \mathrm{Tr}_B[\hat{H}_I(t), [\hat{H}_I(t-\tau), \hat{\rho}_S(t) \otimes \hat{\rho}_B]], \qquad (B.4)$$

where the time evolution is given by the present state $\hat{\rho}_S(t)$ and not on the system past. Hence, no memory effect.

The above procedure is commonly known as the Born–Markov approximation in the literature. In general, it does not guarantee that Eq. (B.4) provides the generator of a dynamical semigroup. Thus, further secular approximation is needed [4]. We proceed by decomposing the interaction Hamiltonian into two parts:

$$\hat{H}_I = \sum_\alpha \hat{A}_\alpha \otimes \hat{B}_\alpha, \qquad (B.5)$$

with $\hat{A}_\alpha^\dagger = \hat{A}_\alpha$ and $\hat{B}_\alpha^\dagger = \hat{B}_\alpha$. The secular approximation is achieved if the interaction Hamiltonian is decomposed in terms of the eigenoperators of the system Hamiltonian $\hat{H}_S$. Let us denote the projection onto the eigenspace belonging to the eigenvalue ϵ in $\hat{H}_S$ as $\hat{\Pi}(\epsilon)$. Then,

$$\hat{A}_\alpha(\omega) = \sum_{\epsilon'-\epsilon=\omega} \hat{\Pi}(\epsilon)\hat{A}_\alpha\hat{\Pi}(\epsilon'). \qquad (B.6)$$

The sum is extended over all energy eigenvalues ϵ' and ϵ of H_S with a fixed energy difference ω. As a consequence, we have $[\hat{H}_S, \hat{A}_\alpha(\omega)] = -\omega\hat{A}_\alpha(\omega); [\hat{H}_S, \hat{A}_\alpha^\dagger(\omega)] = +\omega\hat{A}_\alpha^\dagger(\omega)$. The corresponding interaction picture operators take the form

$$e^{i\hat{H}_S t}\hat{A}_\alpha(\omega)e^{-i\hat{H}_S t} = e^{-i\omega t}\hat{A}_\alpha(\omega), \qquad (B.7)$$

$$e^{i\hat{H}_S t}\hat{A}_\alpha(\omega)e^{-i\hat{H}_S t} = e^{-i\omega t}\hat{A}_\alpha(\omega), \qquad (B.8)$$

with $[\hat{H}_S, \hat{A}_\alpha^\dagger(\omega)\hat{A}_\beta(\omega)] = 0$ and $\hat{A}_\alpha^\dagger(\omega) = \hat{A}_\alpha(-\omega)$. We note that $\hat{A}$'s satisfy the completeness relationship: $\sum_\omega \hat{A}_\alpha(\omega) = \sum_\omega \hat{A}_\alpha^\dagger(\omega) = \hat{A}_\alpha$. Eventually, the interaction Hamiltonian in the interaction picture is then

$$\hat{H}_I(t) = \sum_{\alpha,\omega} e^{-i\omega t}\hat{A}_\alpha(\omega) \otimes \hat{B}_\alpha(t) = \sum_{\alpha,\omega} e^{i\omega t}\hat{A}_\alpha^\dagger(\omega) \otimes \hat{B}_\alpha^\dagger(t), \qquad (B.9)$$

where $\hat{B}_\alpha(t) = e^{i\hat{H}_B t}\hat{B}_\alpha e^{-i\hat{H}_B t}$. By substituting $\hat{H}_I$ back to Eq. (B.4), we arrive at

$$\frac{d\hat{\rho}_S(t)}{dt} = \int_0^\infty d\tau \ \mathrm{Tr}_B \left[\hat{H}_I(t-\tau)\hat{\rho}_S(t)\hat{\rho}_B \hat{H}_I(t) - \hat{H}_I(t)\hat{H}_I(t-\tau)\hat{\rho}_S(t)\hat{\rho}_B \right] + h.c.$$

$$= \sum_{\omega,\omega'} \sum_{\alpha,\beta} e^{i(\omega'-\omega)t} \Gamma_{\alpha\beta}(\omega)(\hat{A}_\beta(\omega)\hat{\rho}_S(t)\hat{A}_\alpha^\dagger(\omega') - \hat{A}_\alpha^\dagger(\omega')\hat{A}_\beta(\omega)\hat{\rho}_S(t)) + h.c.,$$

$$\text{(B.10)}$$

with a bath correlation function

$$\Gamma_{\alpha\beta}(\omega) = \int_0^\infty d\tau e^{i\omega\tau} \langle \hat{B}_\alpha^\dagger(\tau)\hat{B}_\beta(0)\rangle. \tag{B.11}$$

Typical timescale τ_S for which the system evolves is defined as $|\omega' - \omega|^{-1}$, where $\omega' \neq \omega$. By neglecting the rapidly evolving term $\omega' \neq \omega$ during which $\hat{\rho}_S$ varies appreciably, we have

$$\frac{d\hat{\rho}_S(t)}{dt} = \sum_\omega \sum_{\alpha,\beta} \Gamma_{\alpha\beta}(\omega)(\hat{A}_\beta(\omega)\hat{\rho}_S(t)\hat{A}_\alpha^\dagger(\omega) - \hat{A}_\alpha^\dagger(\omega)\hat{A}_\beta(\omega)\hat{\rho}_S(t)) + h.c.$$

$$\text{(B.12)}$$

B.2 An Ultrastrongly Coupled Light–Matter System Weakly Interacts with Bath

Consider a system with a qubit ultrastrongly coupled to a cavity, which is then coupled to a 1-D transmission line via an interaction between the cavity field X and the momentum quadratures Π_ω of the waveguide field outside the cavity.

$$\hat{H}_I = \epsilon_c \int d\omega \hat{X} \hat{\Pi}_\omega = -\epsilon_c \dot{\hat{\psi}}_{cav} \dot{\hat{\psi}}_{tl}(x=0), \tag{B.13}$$

where

$$\hat{\psi}_{tl}(x,t) = \int d\omega \sqrt{\frac{\hbar}{4\pi\omega c}} \left[\hat{b}(\omega)e^{i(kx-\omega t)} + \hat{b}^\dagger(\omega)e^{-i(kx-\omega t)} \right]. \tag{B.14}$$

Here, c is the capacitance per unit length of the resonator and ϵ_c is the coupling constant. For a single mode cavity,

$$\hat{\psi}_{cav}(t) = \sqrt{\frac{\hbar}{2\omega_0 c_r}} \left(\hat{a}e^{-i\omega_0 t} + \hat{a}^\dagger e^{i\omega_0 t} \right), \tag{B.15}$$

where ω_0 is the resonant frequency and c_r is the cavity/resonator capacitance. Therefore,

$$\hat{H}_I = \underbrace{-\epsilon_c - i\sqrt{\frac{\hbar\omega_0}{2c_r}}(\hat{a} - \hat{a}^\dagger)}_{\hat{X}} \int d\omega \underbrace{-i\sqrt{\frac{\hbar\omega}{4\pi c}}(\hat{b}(\omega) - \hat{b}^\dagger(\omega))}_{\hat{\Pi}_\omega}. \tag{B.16}$$

Define $\alpha_0 = \epsilon_c\sqrt{\frac{\hbar\omega_0}{2c_r}}$. Then, we have

$$\hat{H}_I = -i(\hat{a} - \hat{a}^\dagger)\underbrace{i\int_{-\infty}^{\infty} d\omega\alpha_0\sqrt{\frac{\hbar\omega}{4\pi c}}(\hat{b}(\omega) - \hat{b}^\dagger(\omega))}_{\hat{B}_\alpha(0)}. \tag{B.17}$$

The bath correlation function is then given by

$$\Gamma_\alpha(\omega_{kj}) = \int_{-\infty}^{\infty} d\tau e^{i\omega_{kj}\tau} \langle \hat{B}_\alpha^\dagger(\tau)\hat{B}_\alpha(0)\rangle \tag{B.18}$$

$$= \int d\omega\alpha_0^2\left(\frac{\hbar\omega}{4\pi c}\right)[2\pi\hat{N}(\omega)d_\alpha(\omega)\delta(\omega + \omega_{kj}) + 2\pi(N(\omega) + 1)d_\alpha(\omega)\delta(\omega - \omega_{kj})].$$

$d_\alpha(\omega)$ is the bath spectral density. Thus, we arrive at

$$\dot{\rho}_S(t) = -i[\hat{H}, \hat{\rho}_S(t)]$$
$$+ \sum_{k,j>k} -2\pi\alpha_c^2(\omega_{kj})\hat{N}(-\omega_{kj})d_\alpha(-\omega_{kj})[\langle j|\hat{A}_\alpha^\dagger|k\rangle\langle k|\hat{A}_\alpha|j\rangle|j\rangle\langle k|\hat{\rho}_S(t)|k\rangle\langle j|$$
$$-\frac{1}{2}\langle k|\hat{A}_\alpha|j\rangle\langle j|\hat{A}_\alpha^\dagger|k\rangle|k\rangle\langle k|\hat{\rho}_S(t) - \frac{1}{2}\hat{\rho}_S(t)|k\rangle\langle k|\langle k|\hat{A}_\alpha|j\rangle\langle j|\hat{A}_\alpha^\dagger|k\rangle]$$
$$+ \sum_{j,k>j} 2\pi\alpha_c^2(\omega_{kj})(\hat{N}(\omega_{kj}) + 1)d_\alpha(\omega_{kj})[\langle j|\hat{A}_\alpha|k\rangle\langle k|\hat{A}_\alpha^\dagger|j\rangle|j\rangle\langle k|\hat{\rho}_S(t)|k\rangle\langle j|$$
$$-\frac{1}{2}\langle k|\hat{A}_\alpha^\dagger|j\rangle\langle j|\hat{A}_\alpha|k\rangle|k\rangle\langle k|\hat{\rho}_S(t) - \frac{1}{2}\hat{\rho}_S(t)|k\rangle\langle k|\langle k|\hat{A}_\alpha^\dagger|j\rangle\langle j|\hat{A}_\alpha|k\rangle] \tag{B.19}$$

By considering the zero temperature bath, the microscopic master equation reads

$$\dot{\rho}_S(t) = \sum_{j,k>j} i\omega_{kj}\langle j|\hat{\rho}_S(t)|k\rangle|j\rangle\langle k| \tag{B.20}$$

$$+ \sum_\alpha \sum_{j,k>j} \Gamma_\alpha^{jk}\left(|j\rangle\langle k|\hat{\rho}_S(t)|k\rangle\langle j| - \frac{1}{2}\left(|k\rangle\langle k|\hat{\rho}_S(t) + \hat{\rho}_S(t)|k\rangle\langle k|\right)\right),$$

where $\Gamma_\alpha^{kj} = \gamma_\alpha\frac{\omega_{kj}}{\omega_0}|C_{jk}^\alpha|^2$ with $C_{jk} = -i\langle j|\hat{A}_\alpha|k\rangle$, $\hat{A}_\alpha = (\hat{a} + \hat{a}^\dagger)$, $\hat{\sigma}_x$, $\hat{\sigma}_y$, and $\hat{\sigma}_z$. γ_α's are standard damping rates of the weak light–matter coupling regime [5].

Appendix C
Derivation of the Effective Hamiltonian via the Schrieffer–Wolff Transformation

C.1 Schrieffer-Wolff transformation

Let us imagine a system with Hamiltonian H^0 that has two energy subspaces: high-energy and low-energy ones. Now, these two subspaces are weakly coupled by an additional small perturbative term ϵV. However, this perturbation is too small compared to the energy gap Δ present in between the two subspaces. (For the precise description of the Schrieffer–Wolff transformation in quantum many-body systems, the reader is referred to Bravyi et al. [6]). Mathematically,

$$|\epsilon| \leq \epsilon_c = \frac{\Delta}{2||V||}. \tag{C.1}$$

The system we consider has the expression

$$H = H^0 + \epsilon V. \tag{C.2}$$

If we choose to work on the basis of H^0, then it is in diagonal form. However, ϵV is not. In general, ϵV can be decomposed into two parts: one purely block diagonal part H^1 and one purely block off-diagonal part H^2 (see Fig. C.1). Hence, H^2 tries to couple the two energy subspaces according to the strength ϵ.

The purpose of Schrieffer–Wolff transformation is, by going into some generic rotational frame, to diminish the already small coupling between the two subspaces, which is represented by the existence of the block off-diagonal terms, up to a user's chosen order of ϵ by enforcing the following condition:

$$H_{\text{eff}}^{\text{od}} = \sum_{j=1}^{\infty} \frac{1}{(2j+1)!}[H^0 + H^1, S]^{(2j+1)} + \sum_{j=0}^{\infty} \frac{1}{(2j)!}[H^2, S]^{(2j)} = 0, \tag{C.3}$$

© Springer Nature Switzerland AG 2019

T. H. Kyaw, *Towards a Scalable Quantum Computing Platform in the Ultrastrong Coupling Regime*, Springer Theses, https://doi.org/10.1007/978-3-030-19658-5

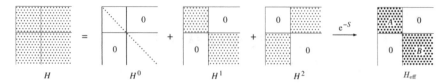

Fig. C.1 The original Hamiltonian consists of an unperturbed diagonal part H^0 and a perturbation that can be decomposed into a block diagonal term H^1 and a block off-diagonal term H^2. The coupling between the blocks A and B is eliminated via the unitary transformation e^{-S} (the Schrieffer–Wolff transformation), yielding a block diagonal effective Hamiltonian. (Reproduced with permission from Ref. [7].)

where S is the frame that one would like to rotate the system into, so that the above condition is satisfied for the specific order of ϵ. Here, "od" means off-diagonal. The transformation S is unique [6]. Furthermore, S has to be anti-Hermitian and block off-diagonal. There are some observations we like to make on Eq. (C.3). As one can see from the setting described above, $H^0 + H^1$ is block diagonal. Thus, the commutator $[H^0 + H^1, S]$ is definitely nonzero since S is block off-diagonal. However, the leftover from the commutator has to be exactly cancelled off with the block off-diagonal term from the system Hamiltonian H^2, which is the second summation term in Eq. (C.3). The so-called rotation S tries to perform this operation up to the user's defined precision of ϵ. In this manner, the effective Hamiltonian is given by

$$H_{\text{eff}} = e^{-S} H e^{S} = \sum_{j=0}^{\infty} \frac{1}{j!} [H, S]^{(j)}. \tag{C.4}$$

The second equality, in Eq. (C.4), comes from the Campbell–Baker–Hausdorff formula. The notation we adopt here is $[H, S]^{(0)} = H$, $[H, S]^{(1)} = [H, S]$, $[H, S]^{(2)} = [[H, S], S]$, and so on. One final ingredient to figure out S is the expression of S itself. Any S can be expressed in a series expansion of ϵ:

$$S = \sum_{j=0}^{\infty} S^{(j)}, \tag{C.5}$$

where $S^{(0)} = 0$, and $S^{(j)}$ is in jth order of ϵ. We can further proceed by plugging this definition of S into the constraint, Eq. (C.3), to terminate all the block off-diagonal terms up to the order ϵ. Since H^1 and H^2 are first order in ϵ, we can derive the following equations, from Eq. (C.3), defining $S^{(j)}$

$$[H^0, S^{(1)}] = -H^2, \text{ first order in } \epsilon \tag{C.6}$$

$$[H^0, S^{(2)}] = -[H^1, S^{(1)}]. \text{ second order in } \epsilon \text{ and so on.} \tag{C.7}$$

Below, one can see that Eq. (C.6) is used to remove the block off-diagonal part in the effective Hamiltonian. With those equations in hand, we can figure out the expression of H_{eff}, from Eq. (C.4), up to second order in ϵ as

$$
\begin{aligned}
H_{\text{eff}} &= [H, S]^{(0)} + [H, S]^{(1)} + [H, S]^{(2)} \\
&= H^0 + H^1 + H^2 + [(H^0 + H^1 + H^2), (S^{(1)} + S^{(2)})] \\
&\quad + \frac{1}{2}[[(H^0 + H^1 + H^2), (S^{(1)} + S^{(2)})], (S^{(1)} + S^{(2)})] \\
&= H^0 + H^1 + \frac{1}{2}[H^2, S^{(1)}] \text{ (block diagonal only).}
\end{aligned}
\tag{C.8}
$$

What we have done so far is to choose a proper operator S such that all terms that couple the two subspaces disappear to the desired order and the effective Hamiltonian is in block diagonal form. However, one can also apply the technique proposed by Poletto et al. [8] and arrive at the effective Hamiltonian in the fully diagonalized form. For our purpose, we would use the effective Hamiltonian obtained in Eq. (C.8).

C.2   Applying to the Catalytic QRS and Two Qubits System

The system-of-interest here is two qubits coupled with a quantum Rabi system (QRS) via the cavity mode of the QRS (see Chap. 6). Following the same notations we used in the main chapter, we recall the system Hamiltonian is given by

$$
\hat{H}_S = \frac{\omega_p}{2}\hat{\sigma}_z^p + \omega_{\text{cav}}\hat{a}^\dagger\hat{a} + g_p\hat{\sigma}_x^p(\hat{a} + \hat{a}^\dagger) + \frac{\omega_{q1}}{2}\hat{\sigma}_z^1 + \frac{\omega_{q2}}{2}\hat{\sigma}_z^2 + (g_1\hat{\sigma}_x^1 + g_2\hat{\sigma}_x^2)(\hat{a} + \hat{a}^\dagger).
\tag{C.9}
$$

If we rewrite the above equation in the language of the Schrieffer–Wolff transformation, we have $H^0 = \frac{\omega_p}{2}\hat{\sigma}_z^p + \omega_{\text{cav}}\hat{a}^\dagger\hat{a} + g_p\hat{\sigma}_x^p(\hat{a} + \hat{a}^\dagger) + \frac{\omega_{q1}}{2}\hat{\sigma}_z^1 + \frac{\omega_{q2}}{2}\hat{\sigma}_z^2$, $H^2 = (g_1\hat{\sigma}_x^1 + g_2\hat{\sigma}_x^2)(\hat{a} + \hat{a}^\dagger)$, and $H^1 = 0$. When we project the entire equation in the basis of the polariton, we arrive at $H^0 = \sum_j \omega_j|j\rangle\langle j| + \frac{\omega_{q1}}{2}\hat{\sigma}_z^1 + \frac{\omega_{q2}}{2}\hat{\sigma}_z^2$, where $|j\rangle$ are eigenstates of the QRS. And, the interaction Hamiltonian $H^2 = H_I = (g_1\hat{\sigma}_x^1 + g_2\hat{\sigma}_x^2)\sum_{j,k>j}[\chi_{kj}|k\rangle\langle j| + \chi_{jk}|j\rangle\langle k|] = \sum_n\sum_{j,k>j} g_n[\hat{\sigma}_+^n\chi_{kj}|k\rangle\langle j| + \hat{\sigma}_+^n\chi_{jk}|j\rangle\langle k| + \hat{\sigma}_-^n\chi_{kj}|k\rangle\langle j| + \hat{\sigma}_-^n\chi_{jk}|j\rangle\langle k|]$, where $\chi_{kj} = \langle k|(\hat{a} + \hat{a}^\dagger)|j\rangle$. In general, we can write the system Hamiltonian as

$$
\hat{H}_S = H^0 + H^1 + H^2 = \sum_j \left(\omega_j|j\rangle\langle j|\right) + \sum_n \left(\frac{\omega_{qn}}{2}\hat{\sigma}_z^n\right)
\tag{C.10}
$$

$$
+ \sum_n\sum_{j,k>j} g_n[\hat{\sigma}_+^n\chi_{kj}|k\rangle\langle j| + \hat{\sigma}_+^n\chi_{jk}|j\rangle\langle k| + \hat{\sigma}_-^n\chi_{kj}|k\rangle\langle j| + \hat{\sigma}_-^n\chi_{jk}|j\rangle\langle k|].
$$

By knowing that S has to be block off-diagonal and anti-Hermitian, i.e., $S^\dagger = -S$, we guess its expression to be

$$S^{(1)} = \sum_n \sum_{j,k>j} [\alpha_n \hat{\sigma}_+^n \chi_{kj} |k\rangle\langle j| - \beta_n^* \hat{\sigma}_+^n \chi_{jk} |j\rangle\langle k| + \beta_n \hat{\sigma}_-^n \chi_{kj} |k\rangle\langle j| - \alpha_n^* \hat{\sigma}_-^n \chi_{jk} |j\rangle\langle k|].$$

(C.11)

Substituting the above equation into Eq. (C.6), $[H^0, S^{(1)}] = -H_I$, we can solve for the expressions for α's and β's. Finally, we obtain

$$S^{(1)} = \sum_n \sum_{j,k>j} [\frac{-g_n}{\delta_{kj}^n} \hat{\sigma}_+^n \chi_{kj} |k\rangle\langle j| - \frac{g_n}{\Delta_{kj}^n} \hat{\sigma}_+^n \chi_{jk} |j\rangle\langle k| + \frac{g_n}{\Delta_{kj}^n} \hat{\sigma}_-^n \chi_{kj} |k\rangle\langle j| + \frac{g_n}{\delta_{kj}^n} \hat{\sigma}_-^n \chi_{jk} |j\rangle\langle k|].$$

(C.12)

With $S^{(1)}$, we apply the result from the previous section, Eq. (C.8), directly to obtain the effective Hamiltonian. To find $[H^2, S^{(1)}]$, we explicitly evaluate

$$[H^2, S^{(1)}]$$

$$= \left[\sum_n \sum_{j,k>j} g_n [\hat{\sigma}_+^n \chi_{kj} |k\rangle\langle j| + \hat{\sigma}_+^n \chi_{jk} |j\rangle\langle k| + \hat{\sigma}_-^n \chi_{kj} |k\rangle\langle j| + \hat{\sigma}_-^n \chi_{jk} |j\rangle\langle k|], \right.$$

$$\sum_{n'} \sum_{l,m>l} [\frac{-g_{n'}}{\delta_{ml}^{n'}} \hat{\sigma}_+^{n'} \chi_{ml} |m\rangle\langle l| - \frac{g_{n'}}{\Delta_{ml}^{n'}} \hat{\sigma}_+^{n'} \chi_{lm} |l\rangle\langle m| + \frac{g_{n'}}{\Delta_{ml}^{n'}} \hat{\sigma}_-^{n'} \chi_{ml} |m\rangle\langle l|$$

$$\left. + \frac{g_{n'}}{\delta_{ml}^{n'}} \hat{\sigma}_-^{n'} \chi_{lm} |l\rangle\langle m|] \right].$$

After (1) expanding the commutator, (2) assuming terms like $|j\rangle\langle k|$, where $j \neq k$ do not contribute in the dynamics (RWA) and thereby considering terms like $|j\rangle\langle j| \ \forall j$ gives

$$[H^2, S^{(1)}] = \sum_{n,n'} \sum_{j,k>j} g_n g_{n'} |\chi_{jk}|^2 \times \qquad\qquad\text{(C.13)}$$

$$\left[\left\{ \left(\frac{1}{\Delta_{kj}^n} + \frac{1}{\delta_{kj}^n}\right) \hat{\sigma}_z^n \delta_{n,n'} + \left(\frac{1}{\Delta_{kj}^n} - \frac{1}{\delta_{kj}^n}\right) \hat{\sigma}_x^n \hat{\sigma}_x^{n'} \right\} |j\rangle\langle j| \right.$$

$$\left. + \left\{ \left(\frac{1}{\Delta_{kj}^n} + \frac{1}{\delta_{kj}^n}\right) \hat{\sigma}_z^n \delta_{n,n'} + \left(\frac{1}{\delta_{kj}^n} - \frac{1}{\Delta_{kj}^n}\right) \hat{\sigma}_x^{n'} \hat{\sigma}_x^n \right\} |k\rangle\langle k| \right].$$

Therefore, the effective Hamiltonian is given by

$$\hat{H}_{\text{eff}} = \sum_j \left(\omega_j |j\rangle\langle j| \right) + \sum_n \left(\frac{\omega_{qn}}{2} \hat{\sigma}_z^n \right) + \frac{1}{2} [H^2, S^{(1)}] \tag{C.14}$$

$$= H_0 + \frac{1}{2} \sum_{j,k>j} |\chi_{jk}|^2 \times$$

$$\left[\left\{ \sum_{n=1}^{2} g_n^2 \left(\frac{1}{\Delta_{kj}^n} + \frac{1}{\delta_{kj}^n} \right) \hat{\sigma}_z^n + g_1 g_2 \left(\frac{1}{\Delta_{kj}^1} + \frac{1}{\Delta_{kj}^2} - \frac{1}{\delta_{kj}^1} - \frac{1}{\delta_{kj}^2} \right) \hat{\sigma}_x^1 \hat{\sigma}_x^2 \right\} |j\rangle\langle j| + \right.$$

$$\left. \left\{ \sum_{n=1}^{2} g_n^2 \left(\frac{1}{\Delta_{kj}^n} + \frac{1}{\delta_{kj}^n} \right) \hat{\sigma}_z^n + g_1 g_2 \left(\frac{1}{\delta_{kj}^1} + \frac{1}{\delta_{kj}^2} - \frac{1}{\Delta_{kj}^1} - \frac{1}{\Delta_{kj}^2} \right) \hat{\sigma}_x^1 \hat{\sigma}_x^2 \right\} |k\rangle\langle k| \right].$$

When considering the two lowest energy levels of the QRS (i.e., $j = 0, k = 1$), we obtain

$$\hat{H}_{\text{eff}} = H_0 + \frac{1}{2} |\chi_{01}|^2 \left[\hat{S}_{12} \otimes \hat{Z}_p + (\hat{Z}_1 + \hat{Z}_2) \otimes \hat{\mathbb{I}}_p \right], \tag{C.15}$$

where $|\chi_{01}|^2 = |\langle 0|(\hat{a} + \hat{a}^\dagger)|1\rangle|^2$, $\hat{Z}_p = |1\rangle\langle 1| - |0\rangle\langle 0|$, $\hat{S}_{12} = g_1 g_2 \left(1/\delta_{kj}^1 + 1/\delta_{kj}^2 \right.$
$\left. -1/\Delta_{kj}^1 - 1/\Delta_{kj}^2 \right) \hat{\sigma}_x^1 \otimes \hat{\sigma}_x^2$, and $(\hat{Z}_1 + \hat{Z}_2) \otimes \hat{\mathbb{I}}_p = \left[\sum_{n=1}^{2} g_n^2 \left(\frac{1}{\Delta_{10}^n} + \frac{1}{\delta_{10}^n} \right) \hat{\sigma}_z^n \right]$
$\otimes (|1\rangle\langle 1| + |0\rangle\langle 0|)$.

As clearly seen in this section and the previous one, the Schrieffer–Wolff transformation is a generalized version of the perturbation theory [6]. There are many other means to attain the effective Hamiltonian besides the Schrieffer–Wolff transformation. One other way is the time averaging operator method proposed by James and Jerke [9], which is a natural mean to be adopted if one is interested to remove fast-oscillating components of the system Hamiltonian, namely the RWA. We have presented this aspect of attaining the effective Hamiltonian in Chap. 6. The main motivation of the present appendix is to show that the two approaches give rise to identical effective Hamiltonian, up to a constant term.

References

1. Romero G, Ballester D, Wang YM, Scarani V, Solano E (2012) Ultrafast quantum gates in circuit QED. Phys Rev Lett 108:120501
2. Peropadre B, Forn-Díaz P, Solano E, García-Ripoll JJ (2010) Switchable ultrastrong coupling in circuit QED. Phys Rev Lett 105:023601
3. Leib M, Hartmann MJ (2014) Synchronized switching in a Josephson Junction Crystal. Phys Rev Lett 112:223603
4. Breuer H-P, Petruccione F (2002) The theory of open quantum systems. Oxford University Press on Demand
5. Ridolfo A, Leib M, Savasta S, Hartmann MJ (2012) Photon blockade in the ultrastrong coupling regime. Phys Rev Lett 109:193602
6. Bravyi S, DiVincenzo DP, Loss D (2011) Schrieffer-Wolff transformation for quantum many-body systems. Ann Phys 326(10):2793
7. Winkler R (2003) Spin-orbit coupling effects in two-dimensional electron and hole systems, volume 191. Springer Science & Business Media

8. Poletto S, Gambetta JM, Merkel ST, Smolin JA, Chow JM, Córcoles AD, Keefe GA, Rothwell MB, Rozen JR, Abraham DW et al (2012) Entanglement of two superconducting qubits in a waveguide cavity via monochromatic two-photon excitation. Phys Rev Lett 109(24):240505
9. James DF, Jerke J (2007) Effective hamiltonian theory and its applications in quantum information. Can J Phys 85(6):625

Printed in the United States
By Bookmasters